101
Amazing Sights
of the Night Sky

A GUIDED TOUR FOR BEGINNERS

by George Moromisato

Adventure Publications
Cambridge, Minnesota

DEDICATION

To Jorge and Celinda, for helping me see the sky.

ACKNOWLEDGMENTS

When I was in high school, my friends and I watched Carl Sagan's *Cosmos* series. We loved to make fun of his "billions and billions," but we were also thoroughly entranced. Astronomy, even via broadcast TV, is best shared, and I thank my friends for the experience. I thank my many teachers, formal and otherwise, who patiently transmitted knowledge while kindling my imagination. Not least, I thank my beautiful wife with whom I can share this and so many other things.

Cover and book design by Jonathan Norberg

Edited by Brett Ortler

Photo credits:

Atlas Image obtained as part of the Two Micron All Sky Survey (2MASS), a joint project of the University of Massachusetts and the Infrared Processing and Analysis Center/California Institute of Technology, funded by the National Aeronautics and Space Administration and the National Science Foundation **217**, Bill Hughes **136**, Brett Ortler **61**, **65**, Bruce Balick and Jason Alexander (University of Washington), Arsen Hajian (U.S. Naval Observatory), Yervant Terzian (Cornell University), Mario Perinotto (University of Florence), Patrizio Patriarchi (Arcetri Observatory) and NASA/ESA **189**, **210**, **219**, **229**, Dan Pelzel **196**, **197**, George Moromisato **41**, **60**, **62**, **66**, **68**, **88**, **90**, **100**, **102**, **104**, **113**, **117**, **123**, **124**, **150**, **174**, **176**, **191**, **193**, **201**, **204**, **223**, **224**, **228**, **246**, **247**, **245**, Helder F. Jacinto **84**, **96**, NASA **122**, **142**. NASA/CXC/Rutgers/J.Warren & J.Hughes et al. **128**, NASA/CXC/Univ. of Alabama/K.Wong et al.   (X-ray), ESO/VLT (Optical) **225**, NASA/Donald R. Pettit **192**, NASA/ESA and The Hubble Heritage Team (AURA/STScI) **213**,
Photo credits continued on page 257

10 9 8 7 6 5 4 3 2 1

Copyright 2017 by George Moromisato
Published by Adventure Publications
An imprint of AdventureKEEN
820 Cleveland Street South
Cambridge, Minnesota 55008
(800) 678-7006
www.adventurepublications.net
Printed in China
ISBN: 978-1-59193-557-5; eISBN: 978-1-59193-684-8

TABLE OF CONTENTS

INTRODUCTION

Once, we were all astronomers. Before electric lights banished the Milky Way, and before digital clocks announced the time, and way before GPS satellites let us know exactly where we are on Earth, the night sky was our entertainment, our timepiece, and our compass.

Understanding the night sky was a matter of life or death. The North Star guided wandering hunters back home, and the rising of the constellations determined when planting should start or when the Nile might flood. But above all, the glowing shapes and patterns of the sky evoked the deepest questions: What's out there? What does it all mean? Are we alone in the universe?

Today, thousands of years later, we've gained knowledge our ancestors could never imagine. We've worked out the motion of the stars and planets; we've unraveled the nature of the starry sky without ever visiting it; and we've probed back in time to study the creation of the universe itself! Yet when I look up into the sky, far vaster and more ancient than anyone could have imagined, I'm faced with the same questions: What's out there? What does it all mean? Are we alone in the universe?

Ultimately, astronomy isn't about the stars; it's about us. We are awed by the scale of the universe, thrilled by the beauty of colliding galaxies, and humbled by the work and dedication that went into discovering it all. We don't *need* to be astronomers anymore, but we *want* to be.

ABOUT THIS BOOK

I wrote this book to share my love of astronomy. I've been an amateur astronomer ever since my parents bought me my first telescope—a mail-order 4-inch Astroscan—and I hope to convey just a little bit of the joy I feel when I'm under the stars. This is not a scholarly work, or an exhaustive reference guide; it isn't even a proper beginner's guide. Instead, think of this book as a "greatest hits" collection. I've compiled a list of the 101 most beautiful, most important, and most fascinating sights in astronomy, and the goal of this book is to take you on a guided tour of them all.

To create the list, I started with several hundred astronomical sights from various catalogs. I excluded any Southern Hemisphere objects that are invisible from North America, and any faint objects that are beyond the reach of the average backyard telescope.

Next, I rated each one based on three criteria: **beauty**, **significance**, and **accessibility**. *Beauty* rates the visual appeal of the object from 1 to 5, with 5 being the best. A solar eclipse rates a 5, while an anonymous faint star rates a 1. *Significance* rates the scientific, historic, or cultural importance of the object. Saturn, for example, is highly significant—it is one of the few astronomical objects that an average person could describe. *Accessibility* rates how easy it is to see. The moon, for example, is relatively easy to see.

I then combined the three ratings into a single number, weighing beauty the most and accessibility the least. The result was a rating for each sight, allowing me to order them into a Top 101 list.

Of course, this list is entirely subjective. Different people will have wildly different ideas about what should go in a Top 101 list, and I certainly don't claim that my list is better than anyone else's. But even if you disagree with my selections, I hope this book will inspire you to get out there, observe the night sky, and maybe even create your own list.

OBSERVING THE NIGHT SKY

Many years ago I took a bird-watching tour in the Peruvian Amazon. We hiked through the rainforest at dawn, while our guide spotted birds. All I could see was trees, but he could always find a colorful parakeet or macaw hidden in the lush green canopy. "Look there! Just below that crooked branch," he would say. Once he pointed it out, it was obvious. But without him, we would have seen nothing.

I asked him how he did it. Was his eyesight unusually sharp? No, he said, it was just experience. After spending years in the rainforest, he could guess where birds were likely to be. He taught me a few tricks, such as always looking at the tallest tree around because birds like to perch there. A few simple guidelines allowed me to see the invisible.

THE MILKY WAY

Observing the many beautiful objects of the night sky is both easier and harder than bird-watching. On the one hand, most sights follow a predictable pattern: the Orion Nebula is always in the sky in winter (at least in the Northern Hemisphere). On the other hand, seeing detail in a faint patch of light will challenge your skills even if you know where to look.

Only experience—hundreds of nights spent under the stars—will turn you into an expert. And while this book is not designed to be a complete field guide, I hope to give you a few tricks to get you started.

THE MOTION OF THE STARS

If you could stand perfectly still out in space, the stars would not move at all—not in less than a human lifespan, anyway. But of course, we're not standing still. We're standing on a planet that spins on its axis once a day and orbits around its sun once a year. We're constantly moving, and therefore the stars appear to move.

Imagine that you're standing on a spinning carousel, looking up at a sky dotted with clouds. As you spin around, you might notice that all the clouds seem to be spinning around a single cloud in the center of your view. There's nothing special about that cloud: it just happens to be directly above the carousel's spinning axis.

STAR TRAILS AND THE NORTH STAR

In the same way, the axis of the world's rotation points toward Polaris—the North Star (see page 172). As the Earth turns, all the stars appear to circle the North Star.

The Earth takes approximately 24 hours to make a full rotation. Since there are 360 degrees in a full circle, the stars around Polaris move 15 degrees per hour. Each hour, some stars will set in the west while others will rise in the east. You'll see a different set of stars at 9 p.m. than at 3 a.m.

At the same time, the Earth is orbiting the sun. In winter we're on the other side of the sun when compared to summer. That means the stars of winter are different from the stars of summer. There are 12 months in the year, therefore the stars shift by 30 degrees every month.

The stars in your sky therefore depend on the date, the time, and your position on Earth (specifically, your latitude). The sky charts in this book (and all printed books) are designed for a specific date, time, and latitude.

Planetarium programs such as *Stellarium*, *TheSky*, and *Starry Night* can accurately plot the skies for any date, time, and location you choose.

THE CONSTELLATIONS

Long before TV beamed sitcoms
and crime dramas to our living
rooms at night, people told each
other stories under the stars. And
it doesn't take much imagination to
convert the random pattern of stars
in the sky into familiar animals, heroic
figures, and even gruesome monsters.

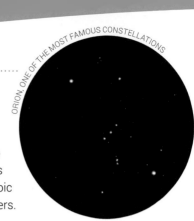

More than 2,000 years ago, Ptolemy codified
48 constellations in his book, *The Almagest*. Many of the most
well-known constellations, like Orion and Leo, were described by
Ptolemy, though they probably predate him by at least a millennium.

Though the ancients did not know it, the stars that make up a constellation are generally unrelated. While they appear close together
from Earth, in reality they could be thousands of light-years apart. For
example, the three stars in Orion's belt appear to be right next to each
other, but in reality, one is twice as far away as the other two.

In 1603, Johann Bayer assigned Greek letters to the brightest stars in
each constellation in descending order of brightness. The brightest
star was named Alpha, the second brightest was named Beta, and so
on. Thus stars could be named by letter and constellation name: Alpha
Centauri is the brightest star in the Centaurus constellation.

In later years, astronomers devised constellations for the stars of the
southern skies, and named them after scientific objects that were then
top of mind: Microscopium (the microscope), Horologium (the clock),
and Fornax (the chemical furnace).

Today, the International Astronomical Union (IAU) defines a constellation as a region of the sky, not just an arrangement of stars. Every part of the sky belongs to one (and only one) constellation. The rectilinear boundaries of the constellations look like the boundaries of the Western US. Every star (and even every galaxy) belongs to one of the 88 modern constellations defined by the IAU.

NAVIGATING THE NIGHT SKY

The easiest way to navigate the night sky is, of course, with a (properly aligned) computerized go-to telescope. There is no shame in using technology to make your life easier. But there are also many reasons to learn how to find objects in the sky on your own, not least being the joy of feeling at home among the stars. When you can look up in the night sky and recognize the stars and constellations, you will feel like a native in your home city, not like a lost tourist.

Start by learning the major star patterns and constellations for each season. Use the seasonal sky charts starting on page 30 to orient yourself at night. There are also many good books and apps for learning the night sky. I can recommend Jonathan Poppele's *Night Sky*, and *Turn Left at Orion* by Consolmagno and Davis.

Once you can recognize the major stars, you will be able to find objects by "star hopping." In this book, I'll often direct you to the star that is closest to the sight; in astronomy, stars typically don't have proper names, so I reference their "official" names, which consist of Alpha, Beta, and so on, and these are labeled on the star charts on each page. Of course, if a star has an actual name, like "Alcor," for instance, I'll reference that instead.

THE BIG DIPPER, PART OF THE CONSTELLATION URSA MAJOR

For example, suppose you're trying to find the Ring Nebula (see the chart on page 158). You can start by aiming your telescope or binoculars at Vega—one of the brightest stars in the sky and one of the stars you should learn to recognize.

By consulting the sky chart you should be able to move from Vega to Zeta Lyrae, the star in the nearest corner of the quadrilateral in Lyra. From there you can move to Beta Lyrae, another corner star. Finally, you can move between Beta Lyrae and Gamma Lyrae to find the Ring Nebula.

THE SOLAR SYSTEM

Unlike the stars, the planets and moons of our solar system move in their own cycles. They are not in the same place at the same time each year.

THE PHASES OF THE MOON

First Quarter

Waxing Gibbous

Waxing Crescent

Full

New

Waning Gibbous

Waning Crescent

Third Quarter

THE MOON The nearest celestial object is the moon; it is tens of millions of times closer than even the nearest star, and yet it took more than three days at hypersonic speeds for the Apollo astronauts to reach it.

The moon changes phase as it moves in its orbit because the sun hits it from a different angle. When the sun hits it face on—as seen from Earth—it's full. When it hits at a right angle, it is half full.

The best time to see the features of the moon is when it is around half full. Craters and mountains at the boundary between light and dark (known as the *terminator*) will have long shadows, which make them easy to see. Use binoculars or a telescope, and focus on the terminator.

THE INNER PLANETS Mercury and Venus are closer to the sun than Earth, and from our vantage point, they don't wander too far away from it. When they are closest to Earth, they are in front of the sun, which means we can't see them. But when they are "full"—that is, when sunlight is hitting them face on, they are farthest from Earth and *behind* the sun.

VENUS IN THE EVENING SKY

The best time to see them is when they are at right angles to the sun with respect to Earth. At those times they are visible after sunset or right before sunrise. This is why Venus is known as the Evening (or Morning) Star.

Use the charts on 157 and 99, respectively, to figure out the best times to see Mercury and Venus.

MARS Every two years and two months (or so) Mars lines up such that it is directly opposite the sun as seen from Earth. This is known as *opposition*, and it is the best time to see the Red Planet.

MARS AND ITS RUDDY GLOW

With the naked eye, Mars is easily recognizable by its reddish-orange tint. Through a telescope, high magnification is required to see any detail. To give yourself the best possible view, observe Mars during opposition, when it's visible all night (and closest to Earth); consult the chart on page 83 for upcoming opposition dates.

JUPITER AND SATURN These gas giants rule the outer solar system. Both are visible through at least part of each year. Jupiter is often brighter than any star. While Saturn is slightly dimmer, it too is unmistakable: if you see a steady, unblinking glow, then you've likely spotted a planet instead of a bright star.

AN IMAGE OF JUPITER THROUGH A TELESCOPE

The chart on page 59 lists the location in the sky to see Jupiter in the years ahead.

Saturn's famous rings change over time. Sometimes we see them as a wide oval circling the planet; other times they are a thin line crossing the equator. The chart on page 43 lists the position of Saturn and where to look.

URANUS AND NEPTUNE Though large compared to Earth, both Uranus and Neptune are too far away to be visible with the naked eye. Neither was known before the age of telescopes, and even today Neptune is difficult to find.

Use a good sky chart to locate these giants. With sufficient magnification, you should be able to see their pale disks, confirming them as planets.

DEEP-SKY OBJECTS

Many of the sights in this book are *deep-sky objects*: nebulae, star clusters, and galaxies. Each type of object requires a slightly different observation technique.

DIFFUSE NEBULAE Our galaxy is filled with diffuse nebulae—vast clouds of interstellar gas—large enough to swallow a dozen star systems. Some of these nebulae are stellar nurseries. Newborn stars have condensed out of the gas and their fusion light now illuminates their birth cocoon. From Earth we see these kinds of nebulae as faintly luminous clouds, sometimes embedded with a clutch of brilliant stars. The Lagoon Nebula (page 100) is a beautiful example.

THE LAGOON NEBULA

To observe a diffuse nebula you need dark, moonless skies away from light pollution. Most of the nebulae in this book are large—sometimes as large as the full moon—so they don't require much magnification. Instead, maximize your light-gathering power: Use short (fast) focal ratios and larger aperture telescopes. For details, see the section on rich-field telescopes (page 235).

Like other faint objects, nebulae benefit from using *averted vision*. Look at the nebula out of the corner of your eye, which is much more light-sensitive than the center.

PLANETARY NEBULAE When a sun-like star dies and collapses into a white dwarf, it leaves behind an expanding shell of gas—the last vestiges of its red giant phase. This shell of gas, lit by the cold fires of the white dwarf, appears from Earth as a small, faint spherical cloud. It looks like a ghostly planet wandering the depths of interstellar space.

THE RING NEBULA

Planetary nebulae are relatively bright, but very small. At low magnification they may seem no more than a slightly fuzzy star. Higher magnification brings out more detail, but you'll need steady skies. These objects are often bright enough to punch through moderate light pollution.

The Dumbbell Nebula (page 138) and the Ring Nebula (page 158) are two of the most famous examples of planetary nebulae.

OPEN CLUSTERS Stars are often born in large groups. The diffuse nebulae that serve as stellar nurseries are large enough to give birth to hundreds of stars at a time.

THE PLEIADES

The intense light of these newborn stars pushes out their dusty cocoon, dissipating the nebula and leaving behind a loose cluster of stars: an open cluster.

From Earth, many of these clusters are small and require a telescope to see. But some of the best are visible even to the naked eye, and they are spectacular in binoculars. The Pleiades (page 76) is the most famous example; Messier 44 (page 184) is another beautiful one.

The best way to observe a cluster depends on its size. For those larger than the full moon, binoculars or a rich-field telescope are best (page 235). Smaller clusters require more magnification.

Unlike nebulae, star clusters are mostly unaffected by light pollution.

GLOBULAR CLUSTERS Whereas open clusters are groups of a few hundred stars, globular clusters pack thousands of stars in a tight spherical (hence, "globular") area.

Globular clusters were formed billions of years ago, probably at the same time the galaxy was born, and they consist mostly of ancient stars.

THE HERCULES CLUSTER

The sight of a globular cluster is unmistakable. They are a tight, cluster of stars, so close together that the core is often difficult to resolve. In smaller telescopes, the core of a globular cluster looks like a fuzzy glow, almost like the core of a galaxy. Larger telescopes and higher magnification are required to resolve the center into individual stars.

Globular clusters are best seen with medium or large telescopes. A high focal-ratio Schmidt- or Maksutov-Cassegrain telescope is ideal (page 235).

GALAXIES Galaxies are the most distant objects visible through amateur instruments. Most are millions of light-years away—a statistic that is nearly incomprehensible. They are visible to us on Earth only because they are so large: 50 to 100 thousand light-years across—more than a thousand times larger than diffuse nebulae.

Nevertheless, galaxies are usually difficult to see. Dark skies, free from light pollution, are a must. Even then, do not expect to see much more than a faint smudge. Use a rich-field telescope with a large aperture for best results.

THE ANDROMEDA GALAXY

DIFFERENT WAYS TO OBSERVE

The frontiers of astronomy require billion-dollar equipment: space telescopes, mountaintop observatories, and super-sensitive digital camera arrays. But ancient observers unraveled many mysteries just by looking up, and no amount of technology can replace the thrill of seeing the Milky Way with your own eyes.

These tips will help you get the most out of whatever equipment you have.

ORION IN WINTER

NAKED-EYE OBSERVING

Some sights, like meteor showers and aurorae, are best seen with the naked eye. But even objects like the Pleiades, which are beautiful through a telescope, can be rewarding visually.

To get the most out of naked-eye observing, invest a little effort into learning the brightest stars and major constellations. Being able to recognize Orion, Leo, Cygnus, and Cassiopeia will help to orient you in the sky. It's like learning the streets of your town by noting landmarks.

The more constellations you recognize, the easier it will be to notice when a planet wanders by. And of course, the more of the sky you know, the easier it will be to find some of the objects in this list.

The biggest obstacle you'll face is probably light pollution. The Milky Way is invisible in almost all urban and suburban areas, which means many faint stars will also be invisible. Our National Parks are some of the last areas in the country where you can see the full beauty of the night sky. Don't miss the opportunity to visit them.

Of course, travel is not always convenient, and even from your backyard there is much to see. You'll need to compensate to make the most of your skies:

- Spend time getting your eyes used to the dark. After 30 minutes in darkness, your eyes will adapt and be able to see fainter stars. If you need a light when you're stargazing, use a red flashlight (or a normal one covered with red cellophane). Red light doesn't "reset" your night vision like white light does.

- Avoid moonlit nights if you want to see the stars. The brightness of the moon washes out the rest of the sky.

- Focus on objects straight overhead (the zenith) instead of close to the horizon.

- Take advantage of nights with excellent transparency. Use the Clear Sky Charts site (www.cleardarksky.com/) for a forecast of sky conditions in your area.

Lastly, prepare appropriately for a long night out. Use a lawn chair to lie down and stare up at the sky. Make sure you have plenty of warm clothes or blankets—there's nothing worse than suffering in the cold. And remember that stargazing is a social activity. You'll enjoy the experience much more if you can share it with friends and family.

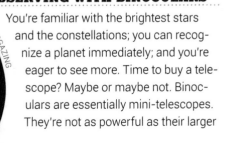

BINOCULARS ARE A GREAT WAY TO START STARGAZING

OBSERVING WITH BINOCULARS

You're familiar with the brightest stars and the constellations; you can recognize a planet immediately; and you're eager to see more. Time to buy a telescope? Maybe or maybe not. Binoculars are essentially mini-telescopes. They're not as powerful as their larger cousins, but they're cheaper, more portable, and much easier to use. Plus you may already have a pair gathering dust in the closet.

There are many sights visible with binoculars, including dozens of nebulae and clusters. Large open clusters like the Hyades, for example, are best in binoculars. Even some large nebulae, like the North America Nebula, are better in binoculars than in high-magnification telescopes.

Consider mounting your binoculars on a tripod. Your hands will get tired holding up even the lightest binoculars. Bring a chair, stool, or even a lawn chair so you can sit comfortably while looking up.

Figure out the field of view of your binoculars. For example, 7 × 50 binoculars can see 6 or 7 degrees of sky—enough to fit 12 or 14 full moons side-by-side. Knowing the field of view will help you to navigate the sky using a chart.

OBSERVING THE SUN

Observing the sun is the only activity in astronomy that presents real danger. Focused sunlight can blind you, and the larger the telescope, the bigger the risk. Just a glance is enough to do permanent damage.

Use front-mounted solar filters from reputable telescope dealers to observe safely. Make sure to leave the lens cap on any optical system without a filter (e.g., a telescope's finder scope). I usually tape down both the filter and any lens caps to prevent accidental exposure.

Another option is *solar image projection*: Mount a pair of binoculars on a tripod *with both lens caps on* and point it at the sun. Use the shadow of the binocular to properly aim. Now remove one of the lens caps and let the image of the sun project down to a piece of white cardboard.

A TELESCOPE WITH A SOLAR FILTER

This method of indirect viewing is great for a group of people. As always, though, make sure curious children don't try to look through the binoculars directly and never leave the setup unattended.

EVEN A SMALL TELESCOPE WILL REVEAL WONDERS

OBSERVING WITH A TELESCOPE

As with binoculars, make sure you know your telescope's field of view. Different eyepieces will provide different magnification (page 236) and thus different fields of view. In general, you'll want to start with low-power eyepieces to find a target and progress to higher powers, if appropriate. For more about telescopes, see Choosing a Telescope (page 235).

Viewing faint objects like galaxies require the best conditions: dark, clear, moonless nights. Start with low power so their faint light is concentrated in a small area—that will make them easier to see.

In addition, faint objects often require an observing technique known as *averted vision*. The centers of our eyes are good at seeing fine detail and color, but not very good in low-light conditions. In contrast, our peripheral vision is sensitive to low light levels but not very good at detail or color. Try looking away from faint objects; you might notice them suddenly appearing bigger and brighter.

Planets, planetary nebulae, and many star clusters can handle higher magnification. These objects generally need steady skies—light pollution doesn't affect them as much. If the object appears to waver as if it's underwater, it means the sky is too turbulent for high magnifications. Wait for a better night or try a different target.

HOW TO USE THIS BOOK

This book lists 101 sights in all, but how you choose to experience them is up to you. A dedicated and experienced amateur might be able to see 90 percent of them in a single night. Rarer events, such as a Great Comet, might take a lifetime to see.

The sights are listed in rank order. The **Top 20** are truly highlights of the sky and are worth every effort to experience. Seeing a total eclipse of the sun, for example, is a rare and unforgettable experience. Many people travel thousands of miles just to see one.

The next 30 sights, though not quite as spectacular, are well worth seeing. They are a cross section of the most important, unusual, or just plain fascinating objects in the sky. Each of these objects is prominent in the history of space and astronomy.

The remaining sights feature everything from far-flung galaxies to spotting Neptune. The Black-Eye Galaxy (page 222), for example, is not the brightest galaxy, but its unusual nucleus and tight arms are far different from those of the Whirlpool Galaxy (page 168). Every sight sparks new questions and highlights the variety of objects in the night sky.

ENTRIES EXPLAINED

BEAUTY Every sight is rated on its beauty from 1 to 5 stars and according to the following scale:

5 stars: Bright and filled with detail or color (e.g., Saturn, Jupiter, or the moon).

4 stars: Bright but small or plain (e.g., Mars); or a bright cluster (e.g., Pleiades); or faint but richly detailed (e.g., Andromeda Galaxy).

3 stars: Bright but featureless (e.g., Mercury); or a sparse cluster; or faint with moderate detail (e.g., M81).

2 stars: Faint and with hard-to-see details (e.g., M78).

1 star: A faint star or smudge, even in a large telescope.

BRAGGING RIGHTS This is the elevator pitch for the object. Why would anyone want to see it (or be proud to have seen it)? It roughly corresponds to the significance component of the rating.

HOW EASY IS IT TO SEE? This describes how accessible an object is and the equipment (if any) required to see it. The following are the most common requirements:

Just look up: The sight is visible to the naked eye.

Best with binoculars or a small telescope: The sight is beautiful in binoculars, but might show additional detail in a small (~4") telescope.

Best with a small telescope: The sight might be visible with binoculars, but a telescope shows it best. Larger telescopes might show additional detail.

Best with a rich-field telescope: The sight requires a telescope with low magnification and a large field of view. Small refractors and Newtonian telescopes are best.

Best with a high-power telescope: The sight requires high magnification. A long focal-length telescope, such as a Maksutov-Cassegrain, is best.

Telescope required: The sight requires a telescope—the bigger, the better.

TYPE The physical category of the sight: a planet, a star, a nebula, etc.

DISCOVERED When the sight was discovered and by whom.

OBSERVING TIPS To get your bearings, be sure to start out by consulting the seasonal star charts on page 30. Each sight includes the best month to view it, as well as rough directions to help you find the object. Each sight also includes a star chart to help you locate the object (when applicable). All the sights have a few observing tips that point out specific details or features to pay attention to. These should help you to get the most out of the observation. If you have trouble finding an object, consult a field guide; it's also very handy to observe with a laptop or a smartphone, as virtual planetarium software can show you exactly where to look.

NOTES The Top 50 sights have a section describing the importance of the sight. It always helps me to appreciate something if I know more about it.

WHAT TO EXPECT When possible, the sight accounts include details about what you'll actually see, including details to make the most of your observation time.

LOCATOR CHARTS Many entries include a locator star chart to help you find the object. These circular charts are aligned so north, toward Polaris, is at the top. The charts are labeled with the compass points; since we're looking up at the sky instead of down at a terrestrial map, east and west are reversed.

WHAT CAN I SEE TONIGHT?

As the Earth revolves around its parent star, different parts of the galaxy appear in the night sky. Most of the 101 sights in this book appear in the same place every year. The following tables organize these sights by the best season to see them.

WINTER

SIGHT	CONSTELLATION
#78 Messier 37	Auriga
#83 Caldwell 5 (IC 342)	Camelopardalis
#34 Sirius	Canis Major
#84 Messier 41	Canis Major
#82 Messier 77	Cetus
#53 Eskimo Nebula (Caldwell 39)	Gemini
#75 Messier 35	Gemini
#63 Rosette Nebula (Caldwell 49)	Monoceros
#67 Christmas Tree Cluster (NGC 2264)	Monoceros
#80 Hubble's Variable Nebula (Caldwell 46)	Monoceros
#14 The Orion Nebula (Messier 42)	Orion
#33 Betelgeuse	Orion
#81 Messier 78	Orion
#32 Algol	Perseus
#36 Double Cluster (Caldwell 14)	Perseus
#91 NGC 1333	Perseus
#93 Messier 46	Puppis
#10 The Pleiades (Messier 45)	Taurus
#23 Crab Nebula (Messier 1)	Taurus
#25 The Hyades (Caldwell 41)	Taurus
#46 Polaris	Ursa Minor

SPRING

SIGHT	CONSTELLATION
#51 Beehive Cluster (Messier 44)	Cancer
#44 Whirlpool Galaxy (Messier 51)	Canes Venatici
#71 Messier 3	Canes Venatici
#74 Messier 106	Canes Venatici
#98 Messier 94	Canes Venatici
#45 Caldwell 38 (NGC 4565)	Coma Berenices
#59 Coma Star Cluster	Coma Berenices
#89 Black-Eye Galaxy (Messier 64)	Coma Berenices
#56 Ghost of Jupiter (Caldwell 59)	Hydra
#69 Messier 83	Hydra
#87 Messier 65 & Messier 66	Leo
#97 NGC 2903	Leo
#92 Caldwell 53 (NGC 3115)	Sextans
#17 Messier 81 And Messier 82	Ursa Major
#43 Virgo Cluster of Galaxies	Virgo
#52 Sombrero Galaxy (Messier 104)	Virgo
#54 Messier 87	Virgo

SUMMER

SIGHT	CONSTELLATION
#35 Albireo	Cygnus
#86 Blinking Planetary (Caldwell 15)	Cygnus
#65 Cat's Eye Nebula (Caldwell 6)	Draco
#19 The Great Hercules Cluster (Messier 13)	Hercules
#90 Messier 92	Hercules
#37 Epsilon Lyrae	Lyra
#39 Ring Nebula (Messier 57)	Lyra
#79 RS Ophiuchi	Ophiuchus
#95 Messier 19	Ophiuchus
#16 Lagoon Nebula (Messier 8)	Sagittarius
#26 Messier 24	Sagittarius
#27 Messier 22	Sagittarius
#47 Swan Nebula (Messier 17)	Sagittarius
#55 Trifid Nebula (Messier 20)	Sagittarius
#58 Messier 6 and Messier 7	Scorpius
#70 Messier 4	Scorpius
#60 Messier 11	Scutum
#20 Eagle Nebula (Messier 16)	Serpens
#62 Messier 5	Serpens
#73 Messier 101	Ursa Major
#29 Dumbbell Nebula (Messier 27)	Vulpecula

FALL

SIGHT	CONSTELLATION
#9 The Andromeda Galaxy (Messier 31)	Andromeda
#96 Caldwell 22 (NGC 7662)	Andromeda
#61 Helix Nebula (Caldwell 63)	Aquarius
#77 Ghost of Saturn (Caldwell 55)	Aquarius
#85 Messier 2	Aquarius
#88 Pac-Man Nebula (NGC 281)	Cassiopeia
#94 Caldwell 13 (NGC 457)	Cassiopeia
#99 Iris Nebula (NGC 7023)	Cepheus
#68 North America Nebula (Caldwell 20)	Cygnus
#72 Veil Nebula (Caldwell 33 and 34)	Cygnus
#76 Messier 15	Pegasus
#100 Messier 74	Pisces
#66 Sculptor Galaxy (NGC 253)	Sculptor
#48 Pinwheel Galaxy (Messier 33)	Triangulum

SPECIAL EVENTS

Of course, many sights don't appear annually in the same constellation; they are not fixed in the sky. The following table lists these special events and notes how often they appear. Note: Planets are best viewed at opposition, but they're visible at other times in the year as well.

SIGHT	HOW OFTEN IS IT VISIBLE
#12 Sunspots and Prominences	Every day
#31 International Space Station	Good opportunity every few days
#8 The Moon and Its Surface	Visible for two or three weeks per month
#13 The Crescent Moon and Earthshine	Every month

#21 Mare Tranquillitatis	Every month
#38 Mercury	Several times a year
#64 Supermoon	Several times a year
#4 Aurorae	A few times a year at northern locations
#50 Zodiacal Light	Spring and fall equinox
#15 Venus and Its Phases	A couple of times a year
#7 The Milky Way	Every year, February to September
#1 Saturn	Best at opposition time, once a year
#41 Uranus	Best at opposition time, once a year
#57 Neptune	Best at opposition time, once a year
#49 Pluto	Best at opposition time, once a year
#30 Fireball Meteor	Unpredictable; perhaps once a year
#40 Titan	Best at Saturn opposition and Titan greatest elongation, generally once a year
#5 Jupiter and Its Moons	Best at opposition time, once every 13 months
#101 Ceres	Best at opposition time, once every 16 months or so
#6 Total Eclipse of the Moon	Every couple of years at a given location
#11 Mars	Best at opposition time, once every 26 months
#42 Planetary Conjunctions and Occultations	Every few years
#18 Partial Eclipse of the Sun	Every few years at a given location
#3 Great Comet	Unpredictable; perhaps once a decade
#22 Venus/Mercury Transit	Once per decade for Mercury transit; once per century for Venus transit
#28 Meteor Storm	Unpredictable; perhaps once a decade
#2 Total Eclipse of the Sun	Every couple of years somewhere on Earth; once a lifetime at a given location
#24 Naked-Eye Supernova	Unpredictable; perhaps once every 300 years

WINTER *(Chart reflects January 1 at 10 p.m. from 40° latitude)*

Despite the cold nights—which you should be prepared for—winter can be one of the best times for observing. The crisp, dry air is unusually transparent and the stars seem to shine brighter. The winter skies are dominated by Orion, one of the easiest constellations to identify and host to several sights on this list. The "W" of Cassiopeia is another prominent constellation; the Milky Way passes through it and you can find dozens of star clusters within it.

#10 PLEIADES (MESSIER 45) ⬡

#14 THE ORION NEBULA (MESSIER 42) ☐

#23 CRAB NEBULA (MESSIER 1) ☐

#25 HYADES (CALDWELL 41) ⬡

#32 ALGOL

#33 BETELGEUSE

#34 SIRIUS

#36 DOUBLE CLUSTER (CALDWELL 14) ⬡

#46 POLARIS

#53 ESKIMO NEBULA (CALDWELL 39) ◇

#63 ROSETTE NEBULA (CALDWELL 49) ☐

#67 CHRISTMAS TREE CLUSTER (NGC 2264) ☐

#75 MESSIER 35 ⬡

#78 MESSIER 37 ⬡

#80 HUBBLE'S VARIABLE NEBULA (CALDWELL 46) ☐

#81 MESSIER 78 ☐

#82 MESSIER 77 ⬭

#83 CALDWELL 5 (IC 342) ⬭

#84 MESSIER 41 ⬡

#91 NGC 1333 ☐

#93 MESSIER 46 ⬡

E

□ = Diffuse Nebula ⬭ = Galaxy ⦂ = Open Cluster ⬦ = Planetary Nebula

31

SPRING *(Chart reflects April 1 at 10 p.m. from 40° latitude)*

As the nights warm up, the Earth faces toward intergalactic space and the realm of galaxies. Many famous galaxies, including the Whirlpool Galaxy, are near the Big Dipper. Though relatively easy to find, you will need dark skies away from light pollution to see them. Leo, the Lion, is also easy to spot, and has its own set of beautiful deep-sky objects.

#17 MESSIER 81 & MESSIER 82 ⬭

#43 VIRGO CLUSTER OF GALAXIES

#44 WHIRLPOOL GALAXY (MESSIER 51) ⬭

#45 CALDWELL 38 (NGC 4565) ⬭

#51 BEEHIVE CLUSTER (MESSIER 44) ⋱

#52 SOMBRERO GALAXY (MESSIER 104) ⬭

#54 MESSIER 87 ⬭

#56 GHOST OF JUPITER (CALDWELL 59) ◇

#59 COMA STAR CLUSTER ⋱

#69 MESSIER 83 ⬭

#71 MESSIER 3 ⊕

#74 MESSIER 106 ⬭

#87 MESSIER 65 & MESSIER 66 ⬭

#89 BLACK-EYE GALAXY (MESSIER 64) ⬭

#92 CALDWELL 53 (NGC 3115) ⬭

#97 NGC 2903 ⬭

#98 MESSIER 94 ⬭

E

SUMMER *(Chart reflects July 1 at 10 p.m. from 40° latitude)*

Earth turns toward the galactic core in summer, treating us to some of the most beautiful objects in the sky. Let the bright, orange star Antares—low in the southern sky—be your guide. It is at the center of the Scorpio constellation and very close to the galactic core. Don't miss the cross representing Cygnus, the Swan, which straddles the Milky Way, and also contains beautiful objects.

#16 LAGOON NEBULA (MESSIER 8) □
#19 THE GREAT HERCULES CLUSTER (MESSIER 13) ⊕
#20 EAGLE NEBULA (MESSIER 16) □
#26 MESSIER 24 ⸪
#27 MESSIER 22 ⊕
#29 DUMBBELL NEBULA (MESSIER 27) ◇
#35 ALBIREO
#37 EPSILON LYRAE
#39 RING NEBULA (MESSIER 57) ◇
#47 SWAN NEBULA (MESSIER 17) □
#55 TRIFID NEBULA (MESSIER 20) □
#58 MESSIER 6 AND MESSIER 7 ⸪
#60 MESSIER 11 ⸪
#62 MESSIER 5 ⊕
#65 CAT'S EYE NEBULA (CALDWELL 6) ◇
#70 MESSIER 4 ⊕
#73 MESSIER 101 ⬭
#79 RS OPHIUCHI -¦-
#86 BLINKING PLANETARY (CALDWELL 15) ◇
#90 MESSIER 92 ⊕
#95 MESSIER 19 ⊕

E

N

65

86 73

37

CYGNUS 39 90

29 LYRA 19

35 HERCULES

AQUILA

60 62

79

20 SCUTUM OPHIUCUS

47

26 95

"the teapot" 55

27 SCORPIUS

SAGITTARIUS 70

16

58

S

□ = Diffuse Nebula ⊕ = Globular Cluster ⬭ = Galaxy ⦂ = Open Cluster
◇ = Planetary Nebula ⬚ = Star Cluster ─|─ = Stellar Object

W

LIBRA

DRACO

FALL *(Chart reflects October 1 at 10 p.m. from 40° latitude)*

The Summer Triangle (Vega, Deneb, and Altair) is lower, but still visible in fall. This area is rich with deep-sky objects. Overhead you should see the Great Square of Pegasus. Though not as bright as some constellations, it is a signpost to two of the most beautiful galaxies in the sky: the Andromeda Galaxy and the Pinwheel Galaxy.

#9 ANDROMEDA GALAXY (MESSIER 31) ⬭

#48 PINWHEEL GALAXY MESSIER 33 ⬭

#61 HELIX NEBULA (CALDWELL 63) ◇

#66 SCULPTOR GALAXY (NGC 253) ⬭

#68 NORTH AMERICA NEBULA (CALDWELL 20) □

#72 VEIL NEBULA (CALDWELL 33 AND 34) □

#76 MESSIER 15 ⊕

#77 GHOST OF SATURN (CALDWELL 55) ◇

#85 MESSIER 2 ⊕

#88 PAC-MAN NEBULA (NGC 281) □

#94 CALDWELL 13 (NGC 457) ⋮

#96 CALDWELL 22 (NGC 7662) ◇

#99 IRIS NEBULA (NGC 7023) □

#100 MESSIER 74 ⬭

E

TOP SIGHTS

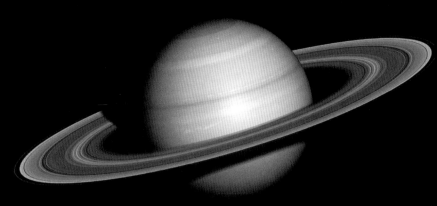

SATURN

· ·

BEAUTY: ★★★★★
BRAGGING RIGHTS: Must see
HOW EASY IS IT TO SEE? Best with small telescope
TYPE: Planet
DISCOVERED: Known since antiquity

NOTES

On any clear night you can look out into space and see some of the planets that share our little solar system. You don't need any special equipment—all you need is to know how to recognize a planet when you see one. There's an easy trick for this: When you see a bright star in the sky, notice whether the star is twinkling or not. If the star is not twinkling, then chances are, you're looking at a planet. That's all there is to it. With that one bit of knowledge you can see the sky in a whole new way, and every time you see a planet you will feel as if you've glimpsed a secret in plain sight.

With even a small telescope, these wandering points of light turn into new worlds. Many planets are visible in the night sky, but one is usually the first target for new stargazers: Saturn! And why shouldn't it be? If you haven't seen Saturn's rings with your own eyes, you're missing out. Seeing a picture of Saturn (even a high-resolution one from a spacecraft) just isn't the same as seeing the real planet through a telescope.

WHAT TO EXPECT

When Galileo looked at Saturn through his telescope, he didn't know what he was seeing! Saturn looked like it had two blobs sticking out the side, like ears. His telescope was too small and crude to see much detail. Your view through binoculars might not be too different. But the higher magnification of even a small modern telescope will start to reveal the rings in all their glory.

WHAT YOU MIGHT SEE THROUGH AMATEUR EQUIPMENT

Obviously, larger telescopes will provide larger, clearer images, but no matter what gear you use, when the planet finally snaps into focus, it's breathtaking.

OBSERVING TIPS

"Planet" means "wanderer." The Big Dipper returns to our winter skies at the same time every year, but planets like Saturn wander through the constellations on their own cycles. Use the chart on the next page to locate Saturn based on the year and season. You'll know you've found it when you see a pale-yellow, unblinking star of moderate brightness. Saturn doesn't shine as brightly as Jupiter or Venus, but it should still be brighter than most stars.

Saturn's rings! As Saturn moves through its orbit, our view of its rings changes. Some years we see them edge-on, as a thin, bright line crossing Saturn's disk. Other years the rings open up and the full beauty of the planet becomes clear. In 2017 the rings will be at one of their widest points, but even on average years the view is magnificent.

At 100× magnification, easily reached with a small telescope, the shadow of the rings on the planet gives the whole view a three-dimensional quality. Seeing Saturn this way for the first time is enough to make you gasp.

Take a closer look at the rings. Though the rings look solid, they are actually composed of billions of icy particles, most no bigger than dust motes, but some as big as a car. The tidal forces of Saturn's many moons herd these tiny particles into complex rings.

At 200× magnification you might find a dark band at the outer edge of the rings. This is called the Cassini Division, named after its discoverer, Giovanni Cassini. It's actually a gap in the rings—the largest of thousands of little gaps.

Now compare the rings on either side of the Cassini Division. Does the inner ring look brighter? This ring has a greater density of icy particles, and thus reflects more light than the others.

Saturn's storm belts. Jupiter's high-contrast storm bands are easier to see, but if you look carefully through a telescope, you can spot subtle bands on Saturn too. At the planet's equator you might see thin little bands, slightly darker than the rest of the disk. You might also notice dark areas at the poles, almost like dark polar caps.

If you haven't seen Saturn's rings with your own eyes, you're missing out.

DATES WHEN SATURN IS AT OPPOSITION
AND WHERE TO LOOK

DATE	CONSTELLATION
June 15, 2017	Ophiuchus
June 27, 2018	Sagittarius
July 9, 2019	Sagittarius
July 20, 2020	Capricornus
August 2, 2021	Capricornus
August 14, 2022	Capricornus
September 8, 2024	Aquarius
September 21, 2025	Pisces
October 4, 2026	Cetus
October 18, 2027	Pisces

TOTAL ECLIPSE
OF THE SUN

BEAUTY: ★★★★★
BRAGGING RIGHTS: Once-in-a-lifetime event
HOW EASY IS IT TO SEE? Requires special equipment
TYPE: Special event
DISCOVERED: Known since antiquity

NOTES

Lunar eclipses happen regularly enough that ancient astronomers were able to work out how to predict them. But total eclipses of the sun, which happen at a given spot only a few times per millennium, were impossible to predict with any accuracy.

Imagine, then, how awesome and scary it must have been for our ancestors to see the life-giving sun swallowed up completely without warning. They must have watched anxiously as the skies darkened, perhaps wondering whether the sun would disappear forever. Within minutes, however, the light and warmth of the sun reappeared and everything would go back to normal—at least for a few more centuries.

Today, of course, we know that the sun is eclipsed when the moon happens to pass in front of it. We can enjoy it as one of nature's greatest spectacles. But it's only the coincidental size and distance of the moon that makes total eclipses possible. If the moon were smaller or further out, it wouldn't be able to cover the sun completely. In fact, the moon is slowly drifting further away from Earth. In a billion years or so, the moon will appear small enough that it won't fully eclipse the sun. The age of total eclipses will be over, so be sure to see a total eclipse of the sun before then.

WHAT TO EXPECT

Seeing a total eclipse of the sun is a once-in-a-lifetime experience. Beyond the rarity of the phenomenon, the visceral feeling of watching the sun disappear and being plunged into darkness will make this an unforgettable event.

WHAT YOU MIGHT SEE THROUGH AMATEUR EQUIPMENT

For North American observers, the best chance to see a total eclipse will come on August 21, 2017, when dozens of US states will witness totality. If you miss that chance, you'll have to wait until 2024, when the moon's shadow will cross from northern Mexico up through the central US and through Maine.

OBSERVING TIPS

Practice safe observing. The most important observing tip is to never look at the sun with unprotected eyes, even when the sun is partially eclipsed. And definitely do not look at it with binoculars or a telescope, as this will cause permanent eye damage. Sunglasses will not prevent this damage; if you want to view it directly, you'll need specially made eclipse glasses, which are available for just a few dollars online. Many astronomy magazines also include such glasses in their issues when eclipses occur.

Project the sun's image. You can use a small telescope, binoculars, or even a pinhole camera to project the sun's image on to a white sheet of paper. This is a perfectly safe way to view the sun and it allows multiple viewers to share the experience. Make sure you aim the instrument without looking through it. See page 20 for instructions.

Observe your surroundings. As the moon begins to eat at the sun, see if you can pinpoint when it starts getting dark. You'll also feel the air get cooler and may even notice animals going into their nighttime routine.

Pay attention to the shadows cast by trees. You may notice hundreds of little crescent suns! The random arrangement of leaves sometimes creates little pinholes that project the sun's image on the ground.

The diamond ring. Just as the moon is about to completely cover up the sun, you'll see one last spot of bright light situated on a bright ring.

This is known as the diamond ring and it's a great moment to photograph, assuming you have a safe setup.

The corona. The highlight of the eclipse will be your view of the corona. With the brightness of the sun's disk obscured, you'll be able to see the superheated gases of the sun's corona streaming away at immense speeds. You'll see a luminous dance of filaments and jets unlike anything you've ever seen.

Until recently, no photograph or video has been able to accurately depict this scene. Unlike our eyes, cameras have trouble capturing both the very bright and very dim features of the corona, so enjoy these precious few minutes.

UPCOMING TOTAL SOLAR ECLIPSES

DATE	VISIBLE FROM
August 21, 2017	Parts of the US, from Oregon to South Carolina
July 2, 2019	Central Argentina and Chile
December 14, 2020	Southern Chile and Argentina
April 8, 2024	Central US, East Canada, Mexico
August 12, 2026	Greenland, Iceland, Spain
August 2, 2027	Morocco, Spain, Algeria, Tunisia, Libya, Egypt, Saudi Arabia, Yemen, Somalia

GREAT COMET

BEAUTY: ★★★★★
BRAGGING RIGHTS: Once-in-a-lifetime event
HOW EASY IS IT TO SEE? Just look up
TYPE: Special event
DISCOVERED: Known since antiquity

NOTES

Comets have always been mysterious. A comet often appears unannounced, spreading its tail across the night sky. Then, a few weeks or months later, it disappears back into the void. Many, such as the famous Halley's Comet, recur on predictable schedules. Others visit once and never return.

Most comets are small and dim, interesting only to amateur astronomers and to the few professionals who study them. But once in a while, a comet is large enough or close enough to attract worldwide attention. The sight of a Great Comet is an unforgettable experience; they are as beautiful as they are rare.

Trillions of icy comets orbit the sun far beyond Pluto—they are so far away that even our best telescopes are unable to track them. Once in a while, a gravitational nudge, perhaps by another star, tips a comet toward the sun.

The comet hurtles toward the inner solar system, slowly at first, but picking up speed as the sun pulls on it. The heat of the sun boils away dust and gases on the comet's surface, causing an enormous tail to form and making it visible from Earth.

Great Comets are rare because they have to fulfill three major requirements: First, they must have a large and active nucleus (capable of emitting lots of gas and dust); second, their orbit must take them close to the sun, so that the sun's energy can heat up the

WHAT YOU MIGHT SEE THROUGH AMATEUR EQUIPMENT

gases; and finally they must pass close to Earth so we can see them in all their glory.

WHAT TO EXPECT

Unfortunately, it's hard to predict when we might see another Great Comet. In the eighteenth century, professional astronomers competed to discover new comets. They searched the skies night after night, hoping to see a fuzzy little comet before anyone else. In recent times, many comets are discovered by amateur astronomers. For example, Yuji Hyakutake discovered a Great Comet, now named after him, in 1996. He used nothing more powerful than a set of binoculars.

Though many lesser comets are found each year, we can expect only one or two Great Comets per decade. As of this writing, the most recent one was Comet McNaught, also known as the Great Comet of 2007. This comet was found by an automated survey in August 2006. By January 2007 it was brighter than Venus and even visible in broad daylight. At night, its tail was larger than the Big Dipper.

OBSERVING TIPS

Watch the comet grow. Comets grow and brighten over time as they get closer to Earth and the sun. Don't wait until they peak to see them. Instead, see if you can spot the comet when it's just a fuzzy little dot. Watch it grow night after night and you'll almost feel as if it's hurtling toward you.

Look for the comet in all conditions. One of my favorite memories of Comet Hale-Bopp (the Great Comet of 1997) is seeing it while walking down the light-polluted streets of Cambridge, Massachusetts. Even though it was only a fuzzy dot, it was still brighter than every other thing in the sky. It hung in the sky like a brilliant cloud.

At their brightest, Great Comets can be seen in broad daylight. They won't be obvious, however. You'll need to figure out where to look, but seeing a comet during the day definitely comes with bragging rights.

One tail or two? Many comets have two visible tails. The first is a bright tail of dust trailing behind the comet's orbital path. But sometimes you can see a second, fainter tail, pointing directly away from the sun. This is a tail of ionized gas that is glowing like a fluorescent bulb, and it is aligned with the sun's magnetic field.

Every tail is different. Pay attention to the tail. Is it broad and bright? Is it long and thin? Is it streaked or uniform? Every Great Comet is different and focusing on the details that make it unique can really help you appreciate it more.

When and Where to Look. Each comet has its own individual orbit around the sun.

Halley's Comet. Halley's Comet is probably the most famous comet of all. Edmond Halley used Newton's then new theory of gravity to determine that a comet that appeared in 1682 was the same as the comet of 1607. He deduced that the comet returned every 76 years and predicted that it would return in 1758. He was right! But he didn't live long enough to see it. Halley's Comet will return in 2061. Perhaps some of you reading this book will see it then.

"I've seen it!"

AURORAE

BEAUTY: ★★★★✦
BRAGGING RIGHTS: Must see
HOW EASY IS IT TO SEE? Dark skies required
TYPE: Special event
DISCOVERED: Known since antiquity

The most important celestial object for us is, of course, the sun. Without it Earth would be a lifeless, frozen hunk of rock; but we take it for granted because it seems constant and unchanging—few things are more certain than "the sun will rise tomorrow."

Aurorae, also known as the Northern Lights, are a visible reminder that the sun does change, and that its behavior can be both predictably cyclical and outright capricious.

In September 1859 a solar storm shot a massive burst of gas and electromagnetic radiation out into space, and Earth just happened to be in its path. The burst induced a massive current on all electrical devices. Fortunately, this was before homes had electric lights, so the only damage was to telegraph equipment, which often burst into flames.

That night, the Northern Lights were visible as far south as Mexico, and people in the US could easily read newspapers by aurora light.

The aurorae occur because of a complex interaction between the solar wind and the Earth's magnetic field. High-energy particles strike the oxygen and nitrogen in our atmosphere, causing them to glow like massive neon signs. And the aurorae become brighter when there's more energy from the sun.

Of course, normal aurorae pose no threat to our cell phones and

WHAT YOU MIGHT SEE THROUGH AMATEUR EQUIPMENT

electronic gadgets—the solar wind is usually far too weak to cause anything more than a spectacular light show. But someday, something like the 1859 solar storm will hit our planet again.

In 2012, a spacecraft in deep space detected a massive burst of energy from the sun. Fortunately, the blast missed the Earth. But based on measurements from the spacecraft, the blast was at least as powerful as the 1859 event. Had it hit us, it would have disabled every unprotected electronic device on the planet. Someday, maybe a century from now, maybe tomorrow, we won't be so lucky. Until then, we can look up and admire the Northern Lights, never forgetting that their gentle beauty is due to the ferocious and unpredictable sun.

WHAT TO EXPECT

The Northern Lights are generally visible in far northern latitudes, from Alaska down to the northern United States. The best time to see them is during the long nights of winter, when there is plenty of darkness. The greater the solar activity, the further south you'll be able to see them.

The sun goes through an 11-year cycle of increased activity followed by a less-active period. Aurorae are most likely to be visible during active parts of the cycle. As of this writing, the most recent cycle (Cycle 24) peaked in early 2014. The next peak is not expected until the mid-2020s.

Unfortunately, there is no guaranteed way to see the Northern Lights, and most predictions of when and where they might be visible are only good a couple of days in advance. Your best bet is to spend an extended period of time at latitudes where aurorae are seen and wait patiently—their beauty will be worth it.

Curtains of light. The aurora appear as multicolored curtains of light waving and dancing in the darkness. They appear somewhat flat or two-dimensional because of the shape of the magnetic field, which channels the high-energy particles along circumpolar arcs.

The gentle waving and pulsing of the aurorae is mesmerizing—clear your mind as you're watching and take in their beauty.

Under the aurorae. Unlike the moon and stars, the aurorae actually occur within our atmosphere. That means you can change your view by getting closer to them. From far away, the aurorae will look like sheets of light near the horizon. But you can travel toward them (perhaps on subsequent nights) to see them better. If you're lucky and travel far enough, you can see them directly above you.

Atomic colors. The colors of the aurorae are caused by excited atoms releasing photons at specific frequencies; this is similar to how particular gases in neon signs produce specific colors of light.

Green is the most common color in aurorae, and it's caused by dense molecular oxygen. Higher in the atmosphere you may see faint reds emitted by more rarefied oxygen. And of course, you may see the aurorae in vivid combinations, sometimes even including yellows and pinks. Molecular nitrogen sometimes plays a part, too, emitting at various wavelengths, but it mostly produces blues and purples.

JUPITER
AND ITS MOONS

BEAUTY: ★★★★⯨
BRAGGING RIGHTS: An amazing sight
HOW EASY IS IT TO SEE? Best with a small telescope
TYPE: Planet
DISCOVERED: Known since antiquity

NOTES

Even 3,000 years ago it was obvious that Jupiter is the King of the Planets. Venus may sometimes shine brighter, but it never strays from the setting or rising sun. In contrast, Jupiter commands the sky, shining brightly throughout the night. It's no wonder the ancient Greeks and Romans named it after their most powerful god.

Today we know that Jupiter is the largest planet in our star system. It is a massive ball of hydrogen, more than twice as heavy as all the rest of the planets combined. An alien astronomer might accurately describe the solar system as consisting of the sun, Jupiter, and assorted other debris.

Jupiter has 67 known moons, forming its own solar system in miniature. Its four largest moons are all bigger than Pluto, and its largest moon, Ganymede, is bigger than Mercury. These four satellites, called the Galilean moons, after their discoverer, Galileo Galilei, are complex and fascinating worlds on their own. Io, the innermost of these, is covered in volcanoes, which are powered by the tidal forces of massive Jupiter. Europa, a little farther out, has a global ocean beneath a layer of ice, and the planet is also kept warm by tidal heating. (It is one of the most likely candidates for life outside of Earth.) Even cold and cratered Callisto is fascinating for its tenuous atmosphere and complex geology.

WHAT YOU MIGHT SEE THROUGH AMATEUR EQUIPMENT

WHAT TO EXPECT

From Earth, Jupiter appears bright, but small. In binoculars you'll just barely be able

to see it as a disk. A telescope with moderate to high magnification will reveal more detail, including the major bands of clouds.

Jupiter is bright enough that you needn't worry about light pollution—even city skies will offer a decent view. But you must wait for crisp and steady skies. Even a little bit of turbulence will blur your view. The best nights to see Jupiter are those cold and clear winter nights when the stars seem bright and steady. If you're using a telescope on a cold night, wait for it to cool down; otherwise, the warm air in the tube will distort your vision, like the ripples rising over hot pavement.

OBSERVING TIPS

Belts and zones. Through a small telescope you will instantly see two parallel dark lines straddling Jupiter's equator. These are the two major equatorial belts. The brighter areas around the belts are known as zones. With higher magnification (and steady skies) you might be able to see other smaller belts above and below the two major ones.

The Great Red Spot. The Great Red Spot is a hurricane large enough to swallow the entire Earth about two times over. It's been raging since at least 1665, when it was discovered by Giovanni Cassini. Despite its name, it may not look very red to you.

Jupiter rotates once every 10 hours, so the Red Spot may not always be facing you. Programs such as Starry Night can predict the position of the Red Spot on any given night. But you can also try observing at different times. If you don't see the Red Spot, try again in two-and-a-half hours, when Jupiter will have turned 90 degrees.

Galilean Moons. The Galilean moons are easy to spot. You'll see one or more little starlike objects on a line crossing Jupiter's equator. The moons move back and forth along the line each night. Galileo plotted their motion from night to night and quickly realized that they orbited

Jupiter. This contradicted the belief that everything orbited the Earth, and lent credence to (though it did not prove) Copernicus's theory of a sun-centered system.

Moon Shadows. Once in a while a Galilean moon is aligned such that it casts a shadow on the face of Jupiter. You'll see a tiny black spot on the disk, moving very slowly across it. Once again, an astronomy program is your best resource for determining when such an event will happen.

DATES WHEN JUPITER IS AT OPPOSITION
AND WHERE TO LOOK

DATE	CONSTELLATION
April 7, 2017	Virgo
May 9, 2018	Libra
June 10, 2019	Ophiuchus
July 14, 2020	Sagittarius
August 19, 2021	Aquarius
September 26, 2022	Cetus
November 3, 2023	Aries
December 7, 2024	Taurus
January 10, 2026	Gemini
February 19, 2027	Leo

TOTAL ECLIPSE
OF THE MOON

BEAUTY: ★★★★✦
BRAGGING RIGHTS: An amazing sight
HOW EASY IS IT TO SEE? Just look up
TYPE: Special event
DISCOVERED: Known since antiquity

NOTES

In contrast to the shock and awe of a solar eclipse, a total eclipse of the moon is a calmer affair. Common enough and consigned to the dead of night, many people are happy to sleep through them. But their beauty should not be underestimated. The moon's ruddy glow, high in the sky, makes the night feel otherworldly, and the unhurried pace gives you a chance to consider and reflect.

You might consider, for example, that this astronomical event is happening on two worlds at the same time: the Earth is seeing a lunar eclipse, but the moon is witnessing a solar eclipse. As you watch the Earth's shadow cross the face of the moon, think about what the moon is seeing.

In 1969, four months after Neil Armstrong's historic journey, the crew of Apollo 12 were on their way back from the moon when their space-ship passed through Earth's shadow. From their perspective, they saw a solar eclipse caused by the Earth! Lunar module pilot Alan Bean reported that "the atmosphere is illuminated completely around the Earth." They could see a dark Earth, surrounded by a ring of light. It was a marvelous sight never before seen by any humans.

WHAT TO EXPECT

Consult the table on the next page for a list of upcoming lunar eclipses visible from North America. The upcoming January 19, 2019, eclipse looks particularly good, with a full hour of totality.

A lunar eclipse involves three distinct phases.

WHAT YOU MIGHT SEE THROUGH AMATEUR EQUIPMENT

It starts with the penumbral phase, in which the moon begins to darken all over. This occurs because the Earth has begun to cover the sun, beginning to darken the face of the moon, but every part of the (sun-facing) moon can still see some bit of the sun.

An hour or more later, you'll see a chunk of the moon disappear. Now we're in the partial phase, when at least a part of the moon is fully obscured by the Earth's shadow.

Over the next hour or so, the moon will gradually disappear until the Earth's shadow covers the entire moon. This is totality, when every part of the moon we can see is experiencing a total eclipse of the sun.

After totality you'll see the phases in reverse, and you'll be left once again with the bright and shiny full moon.

You can watch a lunar eclipse with just your eyes—there's no need for special glasses, binoculars, or telescopes. Nevertheless, the leisurely pace makes this an excellent time to try astrophotography. Even an average telephoto lens can capture the partial or total phases of the eclipse. If you have a telescope, you can usually just put a digital camera right up to the eyepiece to take a photo.

Upcoming Total Lunar Eclipses in North America
- January 31, 2018
- January 21, 2019
- May 16, 2022
- November 8, 2022
- March 14, 2025

OBSERVING TIPS

Observe your surroundings. Take note of your surroundings before the eclipse starts. With the full moon overhead you'll see everything illuminated in bright light. Look again during totality and notice how the reddish color changes your perspective.

Watch the Earth's shadow. In the partial phase you'll see a round bite taken out of the moon—that's the Earth's shadow. Try to imagine how big the whole circle of the shadow must be. That's the size of the Earth compared to the moon.

Totality. Something strange happens toward totality: the dark part of the moon turns red. What's going on? Imagine that you're standing on the moon. You can see that the Earth has completely covered the sun.

But the Earth has an atmosphere! Sunlight from behind the Earth gets bent by its atmosphere and makes its way to the moon. The refracted light is red, just as the sun is when you see it at sunset or sunrise. If you were on the moon, you'd see an Earth surrounded by a ring of fire.

THE MILKY WAY

BEAUTY: ★★★★✦

BRAGGING RIGHTS: You saw our home galaxy

HOW EASY IS IT TO SEE? Dark skies required

BEST TIME TO SEE IT: February to September

TYPE: Galaxy

DISCOVERED: Known since antiquity

NOTES

Some people say that scientific knowledge destroys beauty, that knowing that a rose is just a complex arrangement of carbon-based molecules somehow interferes with appreciating its beauty. But I disagree: appreciating beauty is a visceral feeling, wholly separate from the scientific pleasure of unraveling a puzzle.

So it is with the Milky Way. Once I visited the Grand Canyon at night and saw a brilliant sky, almost as flashy as a Las Vegas casino. And the dominant feature was a frozen river of silver light, arcing across the sky, and bounded by craggy edges. It's a sight that has stayed with me ever since.

Hundreds of years ago, people must have looked at the Milky Way and felt the same thing. Yet today we know something they did not: that the river of light is our home galaxy, a spinning assemblage of 400 billion stars, many hosting planets of their own. Such knowledge only makes the sight more awesome.

WHAT TO EXPECT

Though powered by billions of suns, the Milky Way is so large and sparse that it appears as faint as a ghost to us. The light of the moon overwhelms it, and even a small amount of light pollution renders it completely invisible.

Before cities and electric lights, it was possible to see the Milky Way from anywhere, but today you'll see it only under the darkest skies. I've seen it faintly from

WHAT YOU MIGHT SEE THROUGH AMATEUR EQUIPMENT

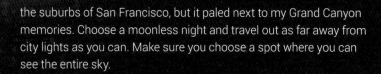

the suburbs of San Francisco, but it paled next to my Grand Canyon memories. Choose a moonless night and travel out as far away from city lights as you can. Make sure you choose a spot where you can see the entire sky.

Once you've found a place, take in the whole view. Just standing beneath the Milky Way is enough to make you fall in love with the night sky. The unaided eye is best to appreciate just how vast it is. A pair of binoculars allows you to resolve the gossamer light into millions of faint stars.

Still, mysteries remain. We do not have a clear view of the galactic core, for example. Vast clouds of dust block all visible light. But by studying other frequencies of radiation—radio waves, infrared, and x-rays—we've been able to determine that a super-massive black hole lives in the heart of our galaxy. Scientists think this black hole is more than 4 million times heavier than our sun. Stars in its vicinity whip around at incredible speeds.

How did this black hole form? How fast is it growing? What's it like to be close to it? We don't know. Future astronomers, perhaps some of you reading this book, will have fun trying to figure that out.

OBSERVING TIPS

The band of light. The Milky Way Galaxy is a flat disk about 100,000 light-years across; the solar system is located about halfway between the galaxy's core and its edge. The band of the Milky Way that we can see is our edge-on view of this galaxy's disk.

Nebulae and star clusters are often found near the band of the Milky Way, since that's where most of the stars in the galaxy are. But galaxies (and many globular clusters) are found away from the band. The Milky Way obscures our view of distant galaxies, so we need to look away from the disk to see them.

Toward the edge. During winter in the Northern Hemisphere, the night side of Earth faces toward the edge of the galaxy (away from the core). Compared to the summer view, you may notice that the band of the Milky Way is a little thinner.

Toward the core. In summer, we face the center of the galaxy. Unfortunately, massive dust clouds block our view, so we can't see the bright core of the galaxy—a sight that would surely be magnificent. Nevertheless, the wide and graceful arc of the Milky Way is splendid. You may notice that in parts the band of the Milky Way is split by darker areas. These are dark dust clouds, obscuring the stars beyond.

The teapot of Sagittarius. The spout of Sagittarius's (see page 35) teapot points to the center of the galaxy. Though we cannot see the actual core, there are marvelous wonders in this region. Use binoculars to scan the area and you'll see star clusters and nebulae just floating in a sea of stars. Many of these objects are included elsewhere in this book, but it's great to see their surroundings as well.

#8

 "I've seen it!"

THE MOON AND ITS SURFACE

BEAUTY: ★★★★★
BRAGGING RIGHTS: A beautiful sight
HOW EASY IS IT TO SEE? Best with binoculars or small telescope
TYPE: Moon
DISCOVERED: Known since antiquity

NOTES

It's easy to take the moon for granted. You're not going to impress someone at a party by saying you saw the moon. But viewing the moon under just a small amount of magnification, with binoculars or a telescope, is enough to make you reconsider. The familiar dark and light splotches on the moon are transformed into a vast desert plain surrounded by fields of overlapping craters.

When you look at the moon you are looking at the surface of another world. Through a telescope you can almost feel that you're flying above the cratered highlands of that alien land. It doesn't take much imagination to see it as a real world that you could walk on, as 12 astronauts once did.

WHAT TO EXPECT

If you use binoculars, make sure you set them up on a tripod so you can take your time looking at all the detail. But a small or medium telescope is preferable, as it helps you zoom in as much as the viewing conditions will allow.

The best time to look at the moon's surface is around the time it is half full. The area between light and dark, called the terminator, is where you'll see the best detail. There, the light from the sun hits the surface at a low angle, causing long shadows that you can see clearly from Earth. All of the hills, craters, and valleys will appear much more prominent at the terminator.

WHAT YOU MIGHT SEE THROUGH AMATEUR EQUIPMENT

The moon is also the easiest object to photograph through a telescope. Once you've pointed your telescope at the moon, all you need to do is point your digital camera (or phone camera) through the eyepiece and just start snapping away. If you can, set the exposure to be 1/60th of a second or faster to keep the image sharp. You may also want to increase the ISO (for more sensitivity) and adjust the white balance (to capture the moon's neutral gray). Practice, as always, will pay off.

OBSERVING TIPS

Maria vs. highlands. The first thing you'll notice is the difference between the darker, flat maria and the rugged and cratered highlands. Ancient astronomers speculated that the flat areas were seas, and named them as such (mare is Latin for sea and maria is the plural form). But today we know that maria are ancient fields of lava.

Early in the moon's history it was bombarded by meteors, which created many of the craters you see. Soon afterwards, volcanic lava flows covered over some of these craters forming the maria.

Tycho crater. One of the most prominent craters is Tycho. Notice how it's surrounded by "rays" that emanate from it and reach far away, even to some nearby maria. These rays consist of ejecta—molten rock that was hurled into the air from the original collision that made the crater. The fact that we can still see the ejecta means that Tycho is a relatively young crater, probably only 100 million years old.

Mare Crisium. Mare Crisium is a small area, notable for being completely surrounded by highlands. The edges of this mare are filled with interesting details: small mountain ridges, half-eroded craters, and even a couple of smaller maria. Since this area is near one edge, you should wait until a couple of days after a full moon for the lighting to be right to observe it.

Clavius crater. It's worth comparing this crater to nearby Tycho. Clavius is much larger, but clearly older, since later impacts have produced craters on top of it. Some craters have obliterated some of Clavius's walls; others have hit right inside.

These are just a few highlights of the lunar surface, but there is so much to see on the moon that we can't fit it all here. Take your time to explore on your own.

"I've seen it!"

THE ANDROMEDA GALAXY (MESSIER 31)

BEAUTY: ★★★★
BRAGGING RIGHTS: It's 2.5 million light-years away!
HOW EASY IS IT TO SEE? Best with binoculars or small telescope
BEST TIME TO SEE IT: Fall (in Andromeda)
TYPE: Galaxy
DISCOVERED: Described in 964 by Abd Al-Rahman Al-Sufi

NOTES

Our galaxy has 400 billion suns! If every person on Earth owned 30 entire star systems, there would still be billions of unclaimed stars in the galaxy. The galaxy is so mind-bogglingly big that when someone suggested that a little wispy cloud in the constellation of Andromeda could be an entire galaxy beyond our own, the reaction was sheer disbelief.

And yet it's true. Look up in the sky on a dark autumn night and find the square of Pegasus (see chart on page 37). Near one of the corners you'll see a very faint patch of light—it'll look like a fuzzy star. Do you see it? You are looking at the combined light of a trillion suns! They are so far away that it took more than 2 million years for their light to reach you.

These numbers are easy to write, and yet entirely incomprehensible. The thrill of seeing the Andromeda Galaxy with your own eyes comes from tackling those numbers. We feel humbled to stand beneath that grandeur.

WHAT TO EXPECT

From Earth, the Andromeda Galaxy is huge; in the sky it appears big as three full moons side by side. Unfortunately, the faint light from this distant galaxy is washed out in all but the darkest skies. If it were a lot brighter, or if our eyes were a lot more sensitive, it would be an awesome sight. With the naked eye you'll only see the bright nucleus as a dim, fuzzy star.

WHAT YOU MIGHT SEE THROUGH AMATEUR EQUIPMENT

The best way to see Andromeda is with binoculars or a low-power telescope under really dark skies. But don't use too much magnification: Andromeda is already big, and too much magnification only dims it more. Instead, try averted vision: look at it out of the corner of your eye and you'll see it brighten.

OBSERVING TIPS

The core of Andromeda. The most obvious feature of Andromeda is its bright core. Here the stars are packed together tightly, like a billion fireflies around a candle flame. All are orbiting a massive black hole heavier than 30 million suns. The center of our own galaxy is very similar, but the Milky Way's nucleus is hidden from us by vast dust clouds. Looking into the heart of Andromeda helps us learn about our own galaxy.

The dark lane. At low power, 25× or below, you might be able to see a dark lane around one side of the nucleus. It will be difficult to see: it's like a sharp boundary in the haze around the nucleus. This is a dust cloud delineating one of the spiral arms of the Andromeda Galaxy.

Unlike the beautiful Whirlpool Galaxy (page 168), which we see face-on, Andromeda is at a shallow angle to us. We see it from the edge, and its spiral structure is difficult to see. Nevertheless, long-exposure photos reveal its shape, and the dark lane is one of its more prominent features.

M32 and M110. Some planets have moons circling them. Many galaxies, including ours, have smaller satellite galaxies orbiting them. Andromeda has two prominent satellites. One, known as Messier 32 (M32), is close to the nucleus and relatively bright. It might look like a star, but if you detect any kind of haze around it, you'll know you've found it. The other, called Messier 110 (M110) is a little farther out and in the opposite direction. It's fainter and harder to see.

Are we alone? The Andromeda Galaxy is so far away that we can't see individual stars with amateur equipment. Instead, a trillion stars blend into a faint haze. Yet it's hard to stare at it and not wonder about all those stars. From studying the Milky Way, we suspect that 1 in 5 sun-like stars have Earth-size planets. Billions of Earth-like worlds could be hidden in that faint haze, invisible to even our best instruments.

Is anyone there staring back? From the Andromeda Galaxy, our own Milky Way would look about the same: as a faint smudge in a dark sky. Perhaps they've looked up with their telescopes and wondered. Regardless, the unimaginable distance between us prevents any kind of communication. Even if we could send a signal, no one there would receive it for more than 2 million years. For now we must be content to just look up and wonder.

THE PLEIADES
(MESSIER 45)

BEAUTY: ★★★★
BRAGGING RIGHTS: You saw the famous Seven Sisters
HOW EASY IS IT TO SEE? Best with binoculars or small telescope
BEST TIME TO SEE IT: Winter (in Taurus)
TYPE: Open cluster
DISCOVERED: Known since antiquity

NOTES

The Pleiades hold a special place in almost every human civilization. To ancient Europeans they were the Seven Sisters; Arab scholars mentioned them in Islamic literature and called them Thurayya; to the Vikings, they were Freyja's hens; and the Japanese named them Subaru, which means "coming together."

On a crisp winter night you can look up and see this beautiful cluster, and perhaps wonder what our ancestors thought as they looked at the exact same sight all those years ago.

Today, thousands of years after they were first observed we know that the Pleiades are a cluster of young and bright stars, all born in the same nebula, and they've only recently (within the last 20 million years) become hot enough to burn off the remnants of their gaseous cocoon. Powerful telescopes reveal far more than the seven traditional members: at least 1,000 stars are thought to be part of the Pleiades.

WHAT TO EXPECT

Faint galaxies and nebulae look wonderful in photos but inevitably disappoint when viewed directly. But bright clusters like the Pleiades are the opposite: photos can't capture the twinkling brightness of these jewels against the black depths of space.

The Pleiades are beautiful even with the naked eye, but a good pair of binoculars or a rich-field telescope will show you dozens more stars. If you use binoculars,

WHAT YOU MIGHT SEE THROUGH AMATEUR EQUIPMENT

mount them on a steady tripod so you can spend time taking in the entire scene.

If you use a telescope, make sure it has a short focal length and low power so you can fit the entire cluster. A 4" reflecting telescope at 25× power is ideal, but a small refractor would also work. Avoid Cassegrain reflectors (such as Schmidt-Cassegrains or Maksutov-Cassegrains), which are great for planets but magnify too much for the Pleiades. (See page 235 for more information.)

OBSERVING TIPS

How many stars can you see? Without using binoculars or telescopes, look up at the Pleiades like our ancestors did. How many stars can you see? There are at least 10 stars brighter than magnitude 6; some are close together and spotting them is a good way to test your vision.

In the days prior to light pollution, stargazers were able to see 12 or 14 stars. On a dark, moonless night away from light pollution, you might be able to match them. Take your time trying to see them all. It might even help to draw a diagram of the stars you see, to better uncover the ones you've missed.

One star or two? With binoculars you'll see dozens of bright stars shining against a black sky. You might notice that some stars that appeared to be solitary are actually two stars close together. Armed with this knowledge, see if you can spot them as doubles without using binoculars.

Nebulosity. Long-exposure photos show the Pleiades enveloped in a faint blue reflection nebula. Though this was once thought to be part of the cluster's original birth nebula, astronomers now believe this is an unrelated nebula that the cluster wandered into. The blue light of these brilliant stars lights up the nebula much like the moon when seen through gossamer clouds.

Seeing the nebula through a telescope will be a challenge. Wait for a moonless night and travel to a place with dark skies—the faintest light pollution will overwhelm this delicate nebula.

But with luck, perseverance, and a rich-field, lower-power telescope, you should be able to see some faint nebulosity around the four brightest stars.

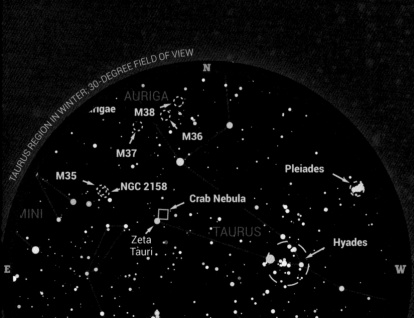

MARS

BEAUTY: ★★★★
BRAGGING RIGHTS: You saw the red planet
HOW EASY IS IT TO SEE? Best with a small telescope
TYPE: Planet
DISCOVERED: Known since antiquity

NOTES

No planet has stirred the human imagination more than Mars. Even in ancient times, the sight of this red star, which some years grows bright and angry, inspired stories of gods and mortals.

Once telescopes were invented, the surface features of this planet, including ice caps and continent-size splotches, inspired scientists and artists alike to speak of Mars as a world like Earth. It was certainly not hard for Percival Lowell, looking at Mars in the 1890s, to imagine that he saw vast canals created by a dying race thirsty for water.

Even after the first space probes visited Mars and saw only barren deserts, the allure of the planet was undiminished. Pictures from the surface, by the Viking landers in the 1970s, made Mars look like any other place on Earth—a place that you could imagine yourself walking around on.

Though Lowell's canals turned out to be imaginary, he was right about one thing: water is the key to life. If there was once water on Mars (and all signs suggest there was, and a lot of it) then perhaps there was life too. And even if there isn't liquid water on the planet now, it exists locked in ice at the frozen poles, where it could easily be used to support the life of future astronauts and even colonizers.

WHAT YOU MIGHT SEE THROUGH AMATEUR EQUIPMENT

Someday, I imagine, maybe in the lifetime of some of you reading this book, humans will land on Mars and walk around. They'll unravel many mysteries, and no doubt stumble into many more.

And maybe bit by bit, they'll turn Mars into something new: no longer a farflung world, but a home.

Until then, and probably even then, we can always look up at Mars and let it inspire us.

WHAT TO EXPECT

The distance from Earth to Mars varies depending on where they are in their orbits. Sometimes Mars is on the other side of the sun from Earth—at those times, not only is Mars invisible (because it is behind the sun) but it is also the farthest away from Earth it can be. Every 26 months or so, Earth and Mars get close to each other. This is known as opposition and it is the best time to view Mars from Earth. The table on the next page lists some upcoming opposition dates.

Even at opposition, however, Mars appears smaller than Jupiter—its tiny disk will be just 25 arcseconds across. (There are 60 arcseconds in an arcminute, and the full moon is 30 arcminutes across, so Mars will appear 72 times smaller than the moon—basically the size of a small lunar crater.)

Use a medium-size telescope to resolve as much detail as you can. Light pollution won't be a problem because Mars is bright, but you must wait for steady skies. Too much turbulence in the air will scramble your view. Cold and clear winter nights are your best bet.

OBSERVING TIPS

Red color. Even without a telescope Mars's color will be obvious. It won't look as saturated as the red light of an airplane's light, but you will notice a distinct reddish/orange hue. This red comes from the oxidized—that is, rusted—sands of Mars.

Retrograde motion. If you draw the position of Mars against the stars you'll notice that it moves from night to night. In the days before opposition, Mars will seem to move backwards. Then after opposition, Mars will start to move forward again. This is known as retrograde motion, and it occurs because of the relative positions of Mars and Earth. As the Earth passes Mars in its orbit, it looks like Mars is temporarily going backwards. Retrograde motion is difficult to explain if all planets revolve around Earth, and it ultimately led to Copernicus's theory that Earth and Mars both revolve around the sun.

Polar ice caps. With sufficient magnification you might be able to see a white patch at one end of Mars. This is a polar ice cap, and it consists of water ice and solid carbon dioxide (dry ice).

Syrtis Major. Seeing other surface features on Mars requires steady skies and lots of patience. Syrtis Major is one of the most obvious. It generally appears as a triangular dark patch. If you're looking through a small telescope, try using a red filter to enhance the contrast. Areas like Syrtis Major will appear darker and hopefully more obvious.

DATES WHEN MARS IS AT OPPOSITION
AND WHERE TO LOOK

DATE	CONSTELLATION
July 27, 2018	Capricorn
October 13, 2020	Pisces
December 8, 2022	Taurus
January 16, 2025	Gemini
February 19, 2027	Leo

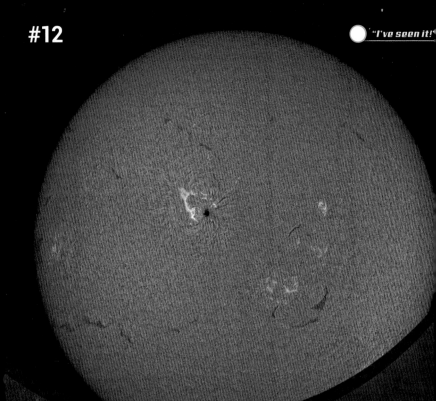

"I've seen it!"

SUNSPOTS
AND PROMINENCES

BEAUTY: ★★★★
BRAGGING RIGHTS: An amazing sight
HOW EASY IS IT TO SEE? Requires special equipment
TYPE: Sun
DISCOVERED: Known since antiquity

The sun is full of mysteries—an unnerving fact given how much we rely on it to sustain all life on Earth. Around 150 years ago, German scientist Julius Mayer calculated that if the sun were a huge ball of burning coal, it would only take a few thousand years for it to burn out. William Thomson—later Lord Kelvin—considered whether energy released by a slowly contracting sun could explain why it shines. He concluded that in such a case it would only last a few million years.

It wasn't until the twentieth century and the notion of nuclear fusion before scientists could begin to explain the sun's prodigious and seemingly inexhaustible output. Yet even then mysteries remained: in the mid-1960s, astronomers measured the number of neutrinos emitted by the sun and discovered far fewer than expected. Since nuclear fusion inevitably produces neutrinos, astronomers wondered whether fusion in the sun's core had decreased or even stopped— a frightful idea. Many years (and one Nobel Prize) later, scientists concluded that the sun was healthy, and that neutrinos were simply morphing into undetectable forms.

Even today, we are far from understanding the sun. In the late nineteenth century, astronomers Annie and Walter Maunder showed that the late 1600s, an unusually cold period in history, coincided with a marked absence of sunspots. Could decreased solar activity have caused the so-called Little Ice Age? And could it happen again?

Sunspots increase and decrease over an 11-year

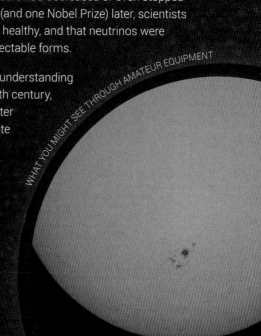

WHAT YOU MIGHT SEE THROUGH AMATEUR EQUIPMENT

cycle, and they are caused by the turbulent electromagnetic forces churning at the heart of the sun. At peak times in this cycle, solar activity increases, causing more solar outbursts and more sunspots.

Yet as the Maunders showed, the sun isn't that simple. Not all cycles are equally active, and sometimes the sun is unusually quiet for many cycles at a time. As of this writing, the current solar cycle, known as Cycle 24, is on track to have the fewest sunspots since modern records began. Will sunspots continue to decrease? Are we heading into another Little Ice Age? We don't know. The sun is full of mysteries.

WHAT TO EXPECT

The sun is almost an ideal observing target. It's visible in the sky every day, barring clouds, and it's easy to find. Unfortunately, it is also the only astronomical object that presents real danger. Looking at the sun directly can harm your eyes and viewing it through unprotected binoculars or a telescope will blind you for life. Do not take any chances when observing the sun, particularly when doing so around children.

To view sunspots and prominences, I recommend purchasing a reputable solar filter, which you can place over the front of your telescope or binoculars. These filters look like mirrored or dark glasses and must be securely and reliably attached before using them. In particular, I'd recommend a hydrogen-alpha filter. These filters are more expensive, but they reveal much more detail than others.

OBSERVING TIPS

Sunspots. Sunspots are areas in which an intense magnetic field has decreased the convection of superheated gases, resulting in cooler— and hence darker—spots. Sunspots have a dark central area, known as the umbra, surrounded by a lighter-colored penumbra.

Prominences. Look at the edge of the sun and you might spot some flame-like structures; these are often looped. These are prominences, and they are vast plumes of superheated gas flowing along a tangled magnetic field.

Chromosphere. A hydrogen-alpha filter allows you to see complex details across the entirety of the solar surface. This is the chromosphere, and it looks like a boiling cauldron of incandescent gas. Like other solar features, the chromosphere is governed by magnetic forces. All the detail you see is caused by the complex rippling magnetic forces of the sun.

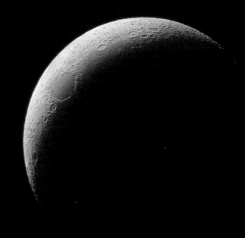

THE CRESCENT MOON AND EARTHSHINE

BEAUTY: ★★★★⟡
BRAGGING RIGHTS: A beautiful sight
HOW EASY IS IT TO SEE? Just look up
TYPE: Moon
DISCOVERED: Known since antiquity

When humans settled down and started planting fields of wheat and rice, the motion of the sun and the stars were used to track time. Solstices and equinoxes announced the passing of the seasons—critical knowledge if your life depended on a bountiful harvest. But before that, when nomadic tribes wandered freely across the plains and steppes, it was the moon that served as the universal clock.

Unlike the sun, the moon changes notably over its 29.5-day cycle. In the first week, the moon grows from a tiny sliver to half full, and each day brings a noticeable change; the moon is full after another week, and then it begins to shrink again. Whereas only high priests could tell you when the solstice happened, even illiterate people could be told to "wait until the next full moon."

It's no wonder that many early calendars were moon-centric, and even now we use months as a primary unit. The Jewish Midrash says that, "the moon has been created for the counting of the days," and in the Islamic calendar, the months traditionally begin when the crescent of a new moon is first sighted.

In our era of atomic clocks and time-synchronized cell phones, we don't need to look at the moon to figure out the date. But it's still beautiful to look at the new crescent moon and watch as it goes through its cycle. Looking at the moon today connects us to a different time, when much of daily life was synchronized to its long orbit.

WHAT YOU MIGHT SEE THROUGH AMATEUR EQUIPMENT

WHAT TO EXPECT

The new moon starts the cycle. The moon is situated in front of the sun and, hence, invisible. At sunset you'll see the crescent moon low on the horizon—it's just a little bit behind the sun and will set soon thereafter.

On the next sunset, the moon will be a little bit higher and the crescent will be bigger and brighter. Each day thereafter, the moon will grow and set later and later. At quarter-full, the moon will be directly overhead when the sun sets. Two weeks from the new moon (or so) the full moon will rise in the east just as the sun is setting in the west.

Day after day the moon will rise later and later. Three weeks after the new moon, a quarter moon will be high in the sky just as the sun rises. Eventually, a crescent moon will hang low in the eastern sky just before sunrise, and then it will be a new moon again.

OBSERVING TIPS

How early can you see it? I love seeing a thin crescent moon set against the fading light of sunset. Somehow it looks closer to us, like it's just a little bit past the horizon.

The closer it is to new moon, the harder it is to see. A very thin crescent is nearly invisible against a bright sky of sunset. But it's easier if you know where (and when) to look. A day after the new moon, start looking at the horizon just after the sun sets.

Earthshine. Each subsequent night the crescent will stay longer and longer past sunset. The skies will grow darker and eventually you'll be able to see that the dark part of the moon is not completely dark! You'll be able to see the whole moon, dim and pale, inside the bright crescent.

This is called Earthshine: the dark part of the moon is being illuminated by the Earth. Anyone on the moon then is seeing a "Full Earth" high overhead, with the bright Earth lighting up the surface.

Watch the moon climb. Each night the moon will set about 50 minutes later than the night before. It will be higher in the sky at sunset. If you watch the moon day after day at the same time, you'll be able to see it move through its cycle, just as our nomadic ancestors once did.

THE ORION NEBULA

BEAUTY: ★★★★✰

BRAGGING RIGHTS: You saw a star factory!

HOW EASY IS IT TO SEE? Best with a small telescope

BEST TIME TO SEE IT: Winter (in Orion)

TYPE: Diffuse Nebula

DISCOVERED: 1611 by Nicolas-Claude Fabri de Peiresc

If you look at Orion on a clear winter night, you might notice that the middle of the three stars in Orion's sword is a little fuzzy. With averted vision, the star seems to grow into a tiny little cloud.

And yet, when Greek astronomers plotted this region of the sky 2,000 years ago, they made no mention of any nebula. The Arab astronomer Al-Sufi noticed the fainter Andromeda "nebula" but made no mention of anything in Orion. Even in 1609, when Galileo looked at Orion's sword with his newly created telescope, he saw nothing but stars. It wasn't until 1611 that Frenchman Fabri de Peiresc finally saw the nebula with his (presumably better) telescope.

Why so many people in the past missed seeing the Great Orion Nebula, a pretty obvious sight for observers today, is a minor mystery in astronomy.

The Orion Nebula, also known as Messier 42 (or M42), is a star factory about 1,300 light-years away. This vast cloud of gas and dust has condensed, via the force of gravity, into newly born stars. Many of these stars are hidden inside dark clouds, but others shine brightly and illuminate the surrounding nebula.

WHAT YOU MIGHT SEE THROUGH AMATEUR EQUIPMENT

In 100,000 years or so, these newborn stars will burn away the surrounding nebula, and they'll shine alone against the black night. By then the Orion Nebula will be no more and we'll see only a bright cluster of stars, perhaps like the Pleiades today. Eventually, the gravitational eddies of the galaxy will break the cluster apart and the stars of the nebula will wander the galaxy alone, just as our sun does now.

WHAT TO EXPECT

The Great Orion Nebula is the best nebula visible from the Northern Hemisphere—only the Southern Hemisphere's Eta Carina Nebula outranks it, and even then I think M42 has a more beautiful and symmetrical shape.

Binoculars or a small telescope are best; large telescopes magnify too much to see the whole nebula. Dark, moonless skies are recommended, but the core of M42 is bright enough to punch through moderate light pollution. No matter where you are, it's probably worth it to give the Orion Nebula a look.

You won't see much, if any, color—nebulae are too faint to excite the eye's color receptors. Fortunately, the Orion Nebula is also bright enough to be an easy target for astrophotography. Other than the moon, there probably is no easier target. With sufficient exposure, you should get some color—mostly pinks and purples—in your astrophotos.

OBSERVING TIPS

Head and wings. The basic shape of the nebula will be obvious on your first view. The heart of the nebula is bright and square, with a small cut, like a mouth. It is flanked by two wings, like those of a ghostly manta ray. Above the heart you'll see a bright knot of nebulosity, technically given its own designation—Messier 43—but it's really just part of the same complex.

Trapezium. With a little bit of magnification, you'll see that the heart of the nebula has four bright stars in a rough trapezoidal shape, hence its colloquial name, Trapezium.

These four stars have already blown away the surrounding nebula gases, which is why we can see them clearly. They also provide the illumination for the rest of the nebula.

With larger telescopes, you might see two additional stars in Trapezium.

How much can you see? The complex structure of the Orion Nebula is beautiful in long-exposure photos, but even with a small telescope you should be able to discern a lot of detail.

Start at the center and try to trace the edges of the nebula. The edges of the wings appear bright, but they fade. How far can you trace them? In really dark skies you might just be able to see that one of the wings wraps almost 180 degrees around the nebula.

VENUS AND ITS PHASES

BEAUTY: ★★★⏵
BRAGGING RIGHTS: You saw the planet closest to the Earth
HOW EASY IS IT TO SEE? Best with a small telescope
TYPE: Planet
DISCOVERED: Known since antiquity

Shining low on the horizon, brighter than any other planet, Venus seems beautiful and peaceful. The Babylonians called it Nin-dar-anna— "Mistress of the Heavens." The Greeks and Romans associated it with the goddess of love. Alas, appearances deceive: the bright clouds of Venus are made of sulfuric acid, and the temperature at its surface is more than 900 °F. Venus is more like hell than paradise.

Early in the twentieth century, scientists imagined Venus as Earth's twin. After all, it is roughly the same size and was known to have a thick atmosphere, unlike the dead moon or cold and deserted Mars. Swedish Nobel prizewinner, Svante Arrhenius, imagined it as a jungle world, hotter than Earth because of its closeness to the sun, but still habitable.

If that had been true, we might be looking for life on Venus instead of Mars. By the 1960s, however, Russian and American space probes revealed the truth: the surface of Venus is hot enough to melt lead. Dreams of living on Venus vanished.

Nevertheless, a puzzle remained: why was Venus so hot? Venus is farther from the sun than Mercury, but it's hotter than it is. Why? In the 1970s, scientists discovered the answer: Venus's thick atmosphere, which is composed mostly of carbon dioxide, traps the heat from the sun, causing a greenhouse effect.

WHAT YOU MIGHT SEE THROUGH AMATEUR EQUIPMENT

Today when you look up at Venus you can see it for what it really is: not a goddess or a jungle world, but a real world. A hostile world, to be sure, but one filled with many mysteries and, possibly, a few lessons for our own planet Earth.

WHAT TO EXPECT

You've probably seen Venus already. Maybe you saw a bright star one day at sunset coming home from work. Or maybe you saw a morning star before dawn. If it wasn't a plane, it was probably Venus.

Venus orbits closer to the sun than the Earth does, which means you'll never see it too far from the sun in the sky. Consult the table on the next page for dates when Venus will be visible.

You'll need binoculars or a small telescope to see the phases of Venus. Though pictures of Venus show cloud patterns, you won't see anything except a bright, featureless disk. Nevertheless, seeing the big crescent of Venus shining in the twilight can be an unforgettable experience.

OBSERVING TIPS

Phases of Venus. From our perspective, we see Venus in one of three positions, depending on where it is in its orbit. When it is on the opposite side of the sun from us, Venus is brightly lit—like a full moon—but it is so far away that we only see a tiny disk.

When Venus is on the same side of the sun as us, it is much closer. Unfortunately, because sunlight is shining behind it, we only see a thin crescent. Nevertheless, because it is so close, the crescent will appear relatively large—even bigger than Jupiter. This is a great time to see Venus.

Finally, Venus can be on one side of its orbit, so it appears half-lit. From our perspective it will appear far from the sun, staying above the horizon long after sunset.

Just like the moon, part of the fun of watching Venus is seeing it pass through its phases night after night. Of course, you'll have to wait longer: whereas the moon completes a cycle in just over 29 days, Venus takes 584 days.

Venus in daytime. Venus is so bright that it is sometimes visible during the day. You just need to know where to look!

Wait for a day when Venus is far to one side of the sun—otherwise it will be lost in the sun's glare. Find Venus just after sunset and try to remember its rough distance away from the sun. The next day, wait for the sun to get lower in the sky and see if you can spot Venus before sunset. Once you get familiar with where Venus is in the sky relative to the sun, it will be easier to find, even in broad daylight.

WHEN TO SPOT VENUS

DATE	WHEN
June 4, 2017	Morning before sunrise
August 18, 2018	Evening after sunset
January 6, 2019	Morning before sunrise
March 25, 2020	Evening after sunset
August 13, 2020	Morning before sunrise
October 30, 2021	Evening after sunset
March 20, 2022	Morning before sunrise
January 4, 2023	Evening after sunset
October 24, 2023	Morning before sunrise
January 10, 2025	Evening after sunset
June 1, 2025	Morning before sunrise
August 16, 2026	Evening after sunset
January 4, 2027	Morning before sunrise

THE LAGOON NEBULA

BEAUTY: ★★★★

BRAGGING RIGHTS: A beautiful sight

HOW EASY IS IT TO SEE? Best with binoculars

BEST TIME TO SEE IT: Summer (in Sagittarius)

TYPE: Diffuse Nebula

DISCOVERED: Before 1654 by Giovanni Hodierna

NOTES

When it comes to astronomy, location matters. Charles Messier observed this object from Paris (49 degrees latitude), which means he saw it obscured by the haze of the horizon. He noted a dim, unimpressive nebula, but otherwise focused on the nearby cluster of stars. He entered both into his famous Messier catalog (see page 253) at number 8.

But from more southern climes, Messier 8—also known as the Lagoon Nebula—is spectacular. It is one of my favorite nebulae, surpassed in North American skies only by the Orion Nebula.

Physically, the Lagoon Nebula is several times larger than the Orion Nebula, but the Orion Nebula is much closer to Earth, so it appears larger and brighter. I can only wonder what such a magnificent nebula would look like if the solar system were closer to it. Once again, location matters.

WHAT TO EXPECT

While not as bright as the Orion Nebula, Messier 8 is still easy to spot. Wait for a moonless night in summer and look for the teapot (page 35) of Sagittarius above the southern horizon. Scan with binoculars above the teapot and you'll see a wealth of wondrous objects: nebulae, clusters, and uncountable stars. The brightest of these will be a small rectangular cloud filled with bright stars—that's the Lagoon Nebula.

WHAT YOU MIGHT SEE THROUGH AMATEUR EQUIPMENT

Dark skies away from light pollution ensure the best view, but you'll be able to see this beautiful nebula from any suburban backyard. Of course, if you're as far north as Maine or Paris, you'll need an unobstructed view of the southern horizon.

You can see plenty of detail with binoculars alone, but a small or medium telescope at low power can reveal much more. Larger telescopes can't always fit the whole nebula, but they reveal more subtle detail at its central core.

As with other faint objects, do not expect to see any color in the nebula.

OBSERVING TIPS

The dark lane. The most obvious feature of the Lagoon nebula is probably the dark lane that splits it into two parts. Agnes M. Clerke wrote about it in *The System of the Stars*, her 1890 popular astronomy book, and referred to the lane as a "lagoon." The name stuck.

With better instruments than those available to her, we can see that this is more like a channel than a lagoon: a thick line of dark dust obscuring the brighter gas beneath. Through a small telescope you can see hints of other dust clouds, threading through the nebula like vines.

The bright folds. The dust clouds give shape to the bright nebula, making it look like folds in satin. Use averted vision to see the full extent of the nebula and to pick up the wealth of detail.

The cluster. Even without a nebula, the tight cluster of bright stars would be beautiful and noteworthy. They are the brightest cluster of stars in Sagittarius and they have their own designation: NGC 6530. Like the Pleiades, these stars were born in the Lagoon Nebula and will eventually drift away, possibly eliminating the rest of the nebula in the process.

The hourglass. The core of the nebula is on the opposite side of the dark lane from the cluster. At medium-high magnification, you might see that the brightest patch is shaped like an hourglass and surrounded by darker nebulae. You're looking at the heart of the nebula, where newborn stars have burned away their dark shroud. The "hourglass" is more like a hole punched through the dark nebula, letting us peer into the star-factory inside.

SAGITTARIUS REGION IN SUMMER, 30-DEGREE FIELD OF VIEW

N

□—— **RS Ophiuchi**

OPHIUCHUS

□—— **Eagle Nebula**

Swan Nebula —□

• **M24**

Trifid Nebula

M21—□

Theta Ophiuchi

M22 —⊕

Antares

Lagoon Nebula (M8) ⊕

M19

E

SCORPIUS

W

SAGITTARIUS

• **M6**

M7

S

MESSIER 81 AND 82

BEAUTY: ★★★★
BRAGGING RIGHTS: A beautiful sight
HOW EASY IS IT TO SEE? Best with a small telescope
BEST TIME TO SEE IT: Spring (in Ursa Major)
TYPE: Galaxy
DISCOVERED: 1774 by Johann Elert Bode

In 1774, amateur-astronomer-turned-professional Johann Bode discovered two nebulous objects close together near the Big Dipper. He had no idea what they were. But neither did anyone else, so Bode added them to his catalog of nebulous objects and carried on. Charles Messier read about them and added them to his own catalog as Messier 81 and Messier 82.

Eventually, Messier 81 (M81) was revealed to be a giant spiral galaxy near our own Milky Way. M82 was also identified as a galaxy, but many mysteries remained. In particular, why did it look like it was exploding? In 1963, astronomers took high-resolution images of M82 and determined that the gaseous filaments seen in the photo were flying out of the center of M82 at more than 500 kilometers per second.

At first astronomers suspected M82 had a violent nucleus, like those seen in quasars and active galaxies like M77 (see page 215). Instead, it turned out Bode's other discovery was to blame: its big brother, M81, was causing M82 to act up.

Millions of years ago, M81 and M82 flew past each other in their long, lazy orbits. The larger galaxy, M81, rattled the smaller one as it flew past, like a speeding truck disturbing a puddle on the side of the road. Density waves—shock waves caused by M81's massive gravity—sloshed inside of M82 and compressed gas near its core. The gas condensed and created thousands and thousands of new stars—a stellar baby

MESSIER 82

boom. Millions of years later, those stars began to reach the end of their lives, and just as they were born together, they died together. Thousands of supernovas exploded in a (relatively) brief period of time, and the collective blast from all those explosions streamed out of the core.

WHAT TO EXPECT

Finding M81 and M82 is not easy; they aren't near any obvious bright star. The easiest way is probably to use the Big Dipper as a guide. Draw a line from Phecda to Dubhe (see chart on next page) and keep following that line for about the width of the Dipper's scoop. Now sweep with binoculars in that area and see if you can spot two fuzzy little objects.

M81 will be brighter and larger; and while M81 appears round, M82 will look elongated, like a ghostly cigar.

While binoculars let you see both objects at once, a small or medium telescope reveals much more detail. As always, dry, clear moonless nights are best for observing faint objects like these.

OBSERVING TIPS

The core of M81. At moderate power you might see some detail in M81's starlike core. Sharp edges in an otherwise smooth glow hint at spiraling dust clouds. Seeing any detail requires practice—a single glance is not enough. The photons from this vast galaxy are being sprayed out to the whole universe, and only a precious few are heading our way.

The arms of M81. Observe M81 at low power and see if you can trace the two spiral arms. They will be very faint and you will need to use averted vision, but persistence will be rewarded. Don't expect to trace them all the way to their tip, as photographs show. Instead, see if you can at least distinguish them from the core.

Low power is essential: you want to concentrate the faint arms into a small area so they are bright enough to see.

M82. You won't see the exploding filaments of M82 without a large telescope, but even at low power on modest instruments you can see some unevenness in this elongated galaxy. Perhaps you'll notice some mottling or segments of unequal brightness.

Averted vision helps again to sense the fainter areas. Try to fit both M81 and M82 in the same field to get good contrast.

PARTIAL ECLIPSE
OF THE SUN

BEAUTY: ★★★★
BRAGGING RIGHTS: A beautiful sight
HOW EASY IS IT TO SEE? Requires special equipment
TYPE: Special event
DISCOVERED: Known since antiquity

Most maps of the solar system are two-dimensional and represent the orbits of the planets by concentric circles. This is a reasonable approximation because planetary orbits are mostly aligned on a plane, known as the ecliptic. Similarly, when you see a diagram of the moon causing eclipses, you see it on a flat plane—the moon goes around the Earth and when it gets in front of the sun, it causes an eclipse. But that poses an obvious question: why don't solar eclipses happen every time?

The reality, of course, is that the real solar system is three-dimensional. The orbit of the moon around the Earth is almost, but not quite, on the same plane as the ecliptic. In truth, the moon's orbit is inclined by five degrees with respect to the Earth's orbit. Most of the time when the moon gets in front of the sun, it is above it or below it in the sky and no eclipse happens.

Eclipses are all about being at the right place at the right time. Imagine the moon's orbit inclined 5 degrees to the ecliptic. The orbit crosses the ecliptic at exactly two points. If the sun happens to be in that direction when the moon is at that exact point, then you get a solar eclipse.

And there's one more factor: the Earth is three-dimen-sional too. Imagine that we're witnessing a total eclipse of the sun. The moon is at the ecliptic and the sun is exactly at the right point. But the Earth is a sphere. Imagine moving north on that sphere: we'll actually be moving above the plane of the ecliptic. If we move far enough, eventually we'll be out of alignment and we won't see a total eclipse.

If everything aligns perfectly, we get a total eclipse. But if we're off by a little bit, then we get a partial one. But even partial eclipses of the sun are rare. The sun and the moon have to align, and you have to be in the right place on Earth at the time it occurs or you won't see anything unusual.

Witnessing even a partial eclipse, we're reminded of the complex clockwork of the solar system. Even ancient astronomers, who kept track of the sky year after year, could not predict solar eclipses with accuracy. It wasn't until we discovered the actual geometry of the solar system, and the exact timing and measurements of the orbits, that we could finally predict the heavens.

WHAT TO EXPECT

List of Upcoming Partial Solar Eclipses

- August 21, 2017: Total in parts of the US; partial in most of North America.

- August 11, 2018: Partial in northeastern Canada.

- June 10, 2021: Annular in parts of Canada; partial in the northeastern US.

- October 14, 2023: Annular in western US; partial in most of North America.

- April 8, 2025: Total in central US; partial in most of North America.

- August 12, 2026: Partial in Canada and northeastern US.

Practice solar observing. Many of the tips pertaining to observing total eclipses (page 44) apply to partial eclipses. Partial eclipses, which happen more often, are a great time to practice your techniques for a full solar eclipse.

Annular eclipses. Even if the moon and the sun are aligned, the moon is sometimes too far away to completely cover the sun. In those cases we get an annular eclipse—the moon leaves a bright ring of sunlight (known as the "ring of fire") behind. Annular eclipses are beautiful for their symmetry, even if you don't get full darkness.

Find the crescent moon. During the day of the eclipse, the new moon and sun are aligned. But the moon will continue to move in its orbit, getting further and further away from the sun. At sunset on the day of the eclipse, see if you can find the very beginnings of the crescent moon. If you can't find it, try again on subsequent days.

"I've seen it!"

THE GREAT HERCULES CLUSTER

BEAUTY: ★★★♩
BRAGGING RIGHTS: An amazing sight
HOW EASY IS IT TO SEE? Best with small telescope
BEST TIME TO SEE IT: Summer (in Hercules)
TYPE: Globular Cluster
DISCOVERED: 1714 by Edmond Halley

I'm genuinely torn. Messier 13, the Great Hercules Cluster, is one of the finest globular clusters in Northern Hemisphere skies—but it is not the finest. That distinction probably belongs to Messier 22, a beautiful globular toward the core of our galaxy. So why is Messier 13 on this list at number 19? As I said in the introduction, the rankings in this book are partly based on beauty and partly based on fame, and the truth is that Messier 13 is much more famous than M22.

Part of that fame comes from its position. While M22 never rises too far above the southern horizon (in North America, anyway), Messier 13 is visible high in the sky, away from all the light pollution. Moreover, while M22 is larger and brighter in small telescopes, bigger telescopes tilt the competition toward M13, which appears full of detail in its compact core.

In 1974, astronomers used the Arecibo radio telescope in Puerto Rico to beam a three-minute encoded message to any possible intelligent life that might live on M13. They chose M13 because it has a high density of stars, and because it was high in the sky at the time.

The message won't reach the Great Hercules Cluster for another 25,000 years or so, and a reply, if any, won't arrive for another 25,000 years after that. But it's still fun to wonder what an alien intelligence, living among the stars of M13, might think about us and our message.

WHAT YOU MIGHT SEE THROUGH AMATEUR EQUIPMENT

Messier 22 may be brighter, but M13, the Great Hercules Cluster, might give you more things to think about.

WHAT TO EXPECT

Messier 13 is relatively easy to find. Look for the constellation Hercules (page 112) and trace out its inner trapezoid of four stars, which make up Hercules's torso. Sweep with binoculars along one side of the trapezoid and you'll see a fuzzy little star. That's the Great Cluster.

To get a better view, use a small telescope at moderate magnification. Bigger telescopes, 8 inches and above, will reveal the core of M13 to be grainy, like a cloud of fireflies from very far away. Use averted vision and the cluster will seem to suddenly grow in size.

OBSERVING TIPS

A half million suns. When you look at Messier 13 you're looking at the combined light of a half million suns. Compare this view with that of the Pleiades. That little cluster is tiny compared to the behemoth you're looking at now, but M13 is so far away—possibly 50 times farther—that its light looks faint and tired by comparison.

Nevertheless, knowing just how far away it is helps you to image how awesome M13 would be if we were closer. Imagine 500 Pleiades clusters packed into a sphere and you might start to get some idea of how it might look.

The heart of M13. Globular clusters are bound together by gravity, with lots of stars in the center and fewer further out. Each star orbits the center of the cluster; together, they form a sphere of stars flying like moths around a flame.

With higher magnification you'll see graininess at the core. Larger telescopes will, of course, show more stars and more detail, but even moderately sized ones are enough to show some features.

NGC 6207. About a moon's width away from M13 you may see a very faint, slightly elongated haze. This is NGC 6207, a distant galaxy that happens to be in the same direction. At magnitude 11, you'll need dark skies and a telescope of 4" or more to see it.

But if you do see it, think about how much farther away the galaxy is. NGC 6207 is likely more than 1300 times farther away than M13. In a single field of view you can get a tiny sense for the incomprehensible vastness of our universe.

EAGLE NEBULA (MESSIER 16)

BEAUTY: ★★★↗

BRAGGING RIGHTS: You saw the "Pillars of Creation"

HOW EASY IS IT TO SEE? Best with a telescope

BEST TIME TO SEE IT: Summer (in Serpens)

TYPE: Diffuse Nebula

DISCOVERED: 1746 by Phillipe Loys de Chéseaux

The frontiers of science often seem far away from us amateurs. None of us is likely to find a cure for cancer in our kitchen or detect the Higgs boson in our garage. And it's unlikely we can afford to launch a space telescope into orbit, even if we save up all our allowance money for the summer.

Yet professional astronomers study the same sky that you can see from your backyard, so whether you're receiving data from the Hubble Space Telescope or looking up with binoculars, the same objects are up there, sending their meager rain of photons to anyone willing to grab them.

Sometimes when I see a magnificent image from Hubble, I remember that I, too, have seen the same object. Ironically, my own faint and hazy view makes the Hubble image come alive. It's like seeing New York City in a movie and remembering the time you visited. Your own memory makes it a real place.

You've probably seen Hubble's famous photo showing columns of glowing gas looming like ghostly stalagmites. They've been dubbed the "Pillars of Creation" because they are creating new stars, but before Hubble they were just part of a faint little nebula in the Serpens constellation, called the Eagle Nebula.

It won't be easy to see this nebula, much less the famous pillars. But whatever you see, I hope you feel a connection to this

WHAT YOU MIGHT SEE THROUGH AMATEUR EQUIPMENT

object that you can't get from the Hubble photo alone. And I hope you feel a connection, however far away, to the frontiers of science.

WHAT TO EXPECT

This is the faintest object in the Top 20. Binoculars reveal the associated cluster of stars, but not the nebula. Even a small telescope is insufficient to see much more than a faint haze.

Wait for a dark and moonless night and use at least an 8-inch telescope —preferably a short-focal length Newtonian.

You can find the Eagle Nebula in the summer. Look for the teapot in Sagittarius and scan the area above it. Above and to the left of the Lagoon Nebula you'll see a ghostly checkmark. That's Messier 17 or the Swan Nebula. Not far above that will be the Eagle Nebula.

At first you'll only see a cluster of stars, but patience (and sufficient aperture) will reveal the nebula.

Don't pass up the chance to see the Eagle Nebula through a large (14-inch-plus) telescope. Star parties or local observatories are a great way to see this elusive, but beautiful, nebula.

OBSERVING TIPS

The nebula. The nebula itself is visible in an 8-inch or larger telescope. The cluster of stars is surrounded by nebulosity: this is the head of the eagle. The "wings" of the eagle are faint patches of nebula spreading out.

Try using a nebula filter, which increases the contrast of the glowing gases. If light pollution is washing out your view, the filter might help.

Pillars. Between the two wings you might be able to see a dark patch of darkness. Scan this area with averted vision. With patience, luck, and maybe a little imagination, you might see one or two pillars of darkness.

Astrophotography. Even modest equipment captures the ghostly pillars. You'll need a telescope with a tracking mount and either a dedicated astronomical camera, or a DSLR with an appropriate adapter. Aim to expose for at least a minute without visible star trails.

Long-exposure astrophotography takes lots of patience and practice. It is not a task for beginners. But once you've perfected your technique on the Orion Nebula, aim your camera at this target and see if you can capture the "Pillars of Creation" for yourself.

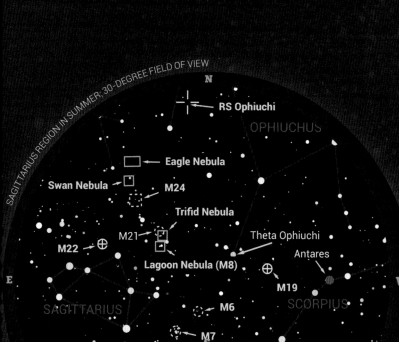

GREAT SIGHTS

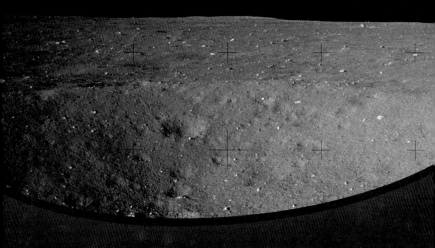

MARE TRANQUILLITATIS
. . . .

BEAUTY: ★★★
BRAGGING RIGHTS: A part of space history
HOW EASY IS IT TO SEE? Best with binoculars or small telescope
TYPE: Lunar feature
DISCOVERED: Known since antiquity; important since 1969

NOTES
. . . .

Mare Tranquillitatis—the Sea of Tranquility—is forever linked with the names Armstrong and Aldrin, yet the first humans to see it up close were Tom Stafford and Gene Cernan. Two months before the historic landing, Stafford and Cernan flew 9 miles above Tranquillitatis in Apollo 10's lunar module.

This "dress rehearsal" tested every part of the moonshot except the landing. The data obtained, and the pilot's view of the landing site, were invaluable to the crew of Apollo 11. And yet, can you imagine traveling 4 days and 250,000 miles only to turn back within 9 miles of a landing and the history books? (Not that Stafford and Cernan had a choice. Their lunar lander—an early design—was too heavy to land on the moon safely.)

When you see Tranquillitatis with your own eyes, it's only right that you think about the Eagle and Armstrong's first words from the surface. But spare a few moments for Stafford and Cernan too.

OBSERVING TIPS

Finding Tranquillitatis. About a week after a new moon you'll see a half moon high in the sky at sunset. A big gray splotch dominates the view, and the center portion of that splotch is the Sea of Tranquility. You can see it with unaided eyes if you know where to look, but binoculars (or a small telescope) give you a more detailed view.

Flat with a chance of boulders. Even at high magnification you can see that Tranquillitatis and the other maria are flat and almost featureless—the perfect place to attempt a risky landing. But appearances from 250,000 miles away can be deceiving. Stafford and Cernan warned that the terrain looked rough within a few miles of the target site. Sure enough, Armstrong had to take manual control during the Apollo 11 mission to avoid crashing into giant boulders.

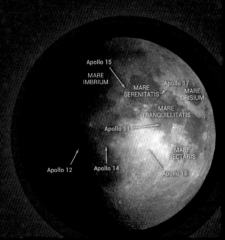

VENUS/MERCURY TRANSIT

BEAUTY: ★★★
BRAGGING RIGHTS: Once-in-a-lifetime event
HOW EASY IS IT TO SEE? Requires special equipment
TYPE: Special event
DISCOVERED: Known since antiquity

NOTES

In the popular imagination, scientists spend their days doing experiments in the lab. There is some truth to this, of course, but what about astronomers? They can't experiment on stars and planets, can they? In fact, they can!

It's true that astronomers can't set up controlled experiments for stars or planets, but they can wait for the cosmos to set it up for them. A transit—when a planet passes in front of the sun—is exactly that: an experiment that Mother Nature has set up for us.

When Mercury passed in front of the sun in 1769, its sharp silhouette revealed that it had no atmosphere. Much rarer transits of Venus were used in the eighteenth and nineteenth centuries to determine the distance from the Earth to the sun.

Today, astronomers look for alien worlds orbiting distant star systems by watching for a slight dip in the star's brightness. If the dimming follows a certain pattern, we deduce that a transit has happened, implying the existence of a planet. We may never be able to visit any of those worlds, but their chance alignment is enough to reveal their existence. All we had to do was look.

OBSERVING TIPS

Transit dates. Transits of Mercury happen about a dozen times per century; the next one will happen in 2019. Transits of Venus are much rarer. The next one won't happen until 2117—another century from now. For detailed transit information, visit NASA's spectacular Eclipse site: http://eclipse.gsfc.nasa.gov/eclipse.html.

Practice safe observing. Use appropriate solar filters to observe the sun. Do not take any chances with your eyesight. Solar projection also works well (see page 20 for details).

Black-drop effect. When the transit is about to occur, look to the sun (again, with eye protection). As Mercury (or Venus) first crosses the edge of the sun, you'll see more and more of the silhouette. Eventually you'll see a black circle just touching the edge of the sun—it will look like a black drop hanging on to the sun's edge. Moments later the drop will detach and the planet will be fully inside the sun's circle as it moves across the face of the sun.

CRAB NEBULA (MESSIER 1)

BEAUTY: ★★★
BRAGGING RIGHTS: You saw a dead star
HOW EASY IS IT TO SEE? Best with a small telescope
BEST TIME TO SEE IT: Winter (in Taurus)
TYPE: Supernova Remnant
DISCOVERED: 1731 by John Bevis

NOTES

Our sun is not heavy enough to explode into a supernova. Once it runs out of hydrogen and helium, the sun's gravity will compress it down to a white dwarf—a compact star the size of the Earth.

But larger stars have a more spectacular fate. A star 10 times heavier than the sun will not stop collapsing at the white dwarf stage. It will continue collapsing, all the while fusing heavier and heavier elements, until it forms a core of solid iron, which cannot be fused efficiently. Then the outer layers crash down on this incompressible core, and the star flares up in an explosion visible across the universe: a supernova.

In the year 1054, Chinese astronomers recorded that titanic event happening in the constellation of Taurus. A new star appeared, outshining Venus and it was visible even in daylight. In time it faded, and after a few years, it was no longer visible. At least not with the naked eye.

But today, if you point your telescope at that exact spot, you'll see the ghostly remains of that explosion. Messier 1 is all that's left of an old cataclysm that, for a brief time, was as bright as our entire galaxy.

OBSERVING TIPS

Dark skies required. You can find the Crab Nebula about 1 degree (two full moon widths) away from Zeta Tauri (see chart below).

The remnant of that long-ago explosion has faded considerably. It shines at magnitude 8 or 9, just possibly visible with binoculars under dark skies. A telescope reveals a fuzzy patch of light, but much patience is required to see any detail.

A ragged shell. Compare your view of M1 against those of the Ring Nebula (page 158) and the Dumbbell Nebula (page 138). Those planetary nebulae look more symmetrical, almost stately, compared to the ragged and misaligned supernova remnant.

The pulsar. The massive neutron core of the star is still there. It holds more weight than our sun in a sphere just 12 miles across—about the length of Manhattan. Unfortunately, the dead star—spinning rapidly and emitting fierce radiation—is invisible in all but the largest amateur telescopes.

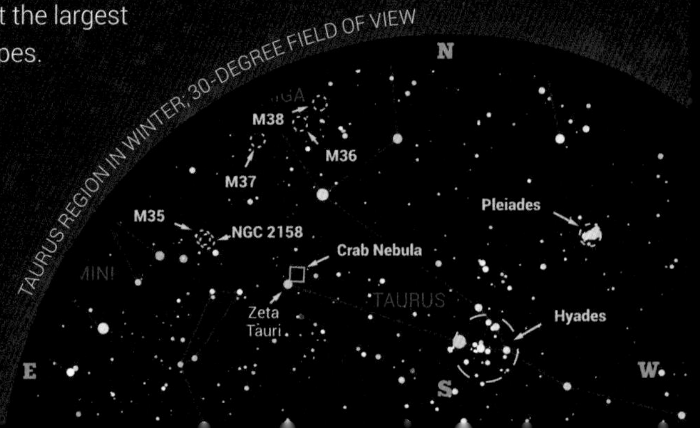

TAURUS REGION IN WINTER; 30-DEGREE FIELD OF VIEW

N

M38

M36

M37

M35

NGC 2158

Pleiades

Crab Nebula

TAURUS

Zeta Tauri

Hyades

E

S

W

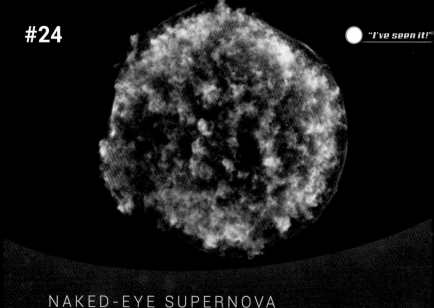

NAKED-EYE SUPERNOVA

BEAUTY: ★★★
BRAGGING RIGHTS: Once-in-a-lifetime event
HOW EASY IS IT TO SEE? Just look up
TYPE: Special event
DISCOVERED: Known since antiquity

NOTES

Do the stars influence us here on Earth? Astrologers would have you believe so—but there is no evidence they're right and plenty of evidence they're wrong. But one evening in 1572, a man named Tycho Brahe was staring at the sky when he noticed a new bright star he did not recognize. It was brighter than Venus and even visible in broad daylight the next day.

For months this nova stella—new star—stayed in the sky while people on Earth argued about its nature. Was it a compact comet? Was it some fluke of the atmosphere? Tycho Brahe did more than speculate. He measured its position carefully and discovered that it didn't move from night to night like a comet. It was therefore likely a phenomenon of the stars—not of the atmosphere or the solar system.

The thrill of seeing a new star and being able to probe its nature inspired Brahe. He devoted his considerable fortune to building an observatory for the measurement and study of the stars. Brahe's measurements (and money) allowed Johannes Kepler to explain and predict the motion of the planets, which led Newton to formulate his laws of motion, knowledge that is used even today to launch missions into space.

Ironically, Brahe was a firm believer in astrology. In those days, the line between science and superstition was thin. But perhaps he also had the excuse that the stars—one new star in particular—did influence him directly.

OBSERVING TIPS

The rarest of the rare. Our galaxy has more than 400 billion suns, and most shine brightly year after year, century after century. Every 50 years or so, one of those billions of stars ends its life in a cataclysmic explosion: a supernova. But most of those explosions are invisible from Earth, hidden behind the massive dust clouds of the galactic disk. Only a relatively nearby supernova will be visible to the naked eye from Earth. Naked-eye supernovas are incredibly rare—possibly as infrequent as one every 300 years. No one alive has seen a supernova in our galaxy in visible light, and it's possible everyone now alive will be dead by the time the next one happens. Or it could happen tomorrow.

See it during the day. There is no danger to Earth from a supernova. These stars are unimaginably far away. A piece of cardboard is enough to shield you from their energy, despite the fact they are gargantuan thermonuclear explosions in space. Nevertheless, they can be spectacular. At its peak, a supernova outshines the rest of its galaxy. From Earth, even a distant supernova should be visible in the day.

Watch it fade. Night after night the supernova will fade, as its energy dissipates into the void. Brahe's nova dimmed and changed colors over time. What will you see? Perhaps—if you're lucky—you'll live long enough to see one.

THE HYADES (CALDWELL 41)

BEAUTY: ★★★⭒
BRAGGING RIGHTS: A beautiful sight
HOW EASY IS IT TO SEE? Best with binoculars
BEST TIME TO SEE IT: Winter (in Taurus)
TYPE: Open Cluster
DISCOVERED: Known since antiquity

NOTES

The Hyades are our stellar neighbors. If the Milky Way were 60 miles across—the size of Los Angeles—the Hyades would be less than 500 feet away from us, literally a few houses down the street. In contrast, the Pleiades (page 76), would be a quarter mile away—almost three times farther.

And yet the Pleiades are a lot brighter. Why? The Pleiades are a young cluster dominated by bright, but short-lived, blue giants. The Hyades are a much older cluster, and its original blue giants have long since collapsed into white dwarfs. The stars that remain are the smaller, sun-like stars that burn for billions of years.

There is something comforting in that. The Hyades are a nice quiet cluster filled with middle-aged stars not too different from our own sun. In contrast, the Pleiades are like Hollywood movie stars: burning hot and bright and beautiful from a distance, but not always the best neighbors.

It's only recently that we've been able to measure the distance to the Hyades. In 1989, the European Space Agency launched the Hipparcos satellite, the first space telescope designed to precisely measure the position, brightness, and distance of thousands of stars, including those of the Hyades.

The turbulence of the atmosphere, which blurs and distorts starlight, prevents us from making very accurate measurements from the ground, even with the largest telescopes. But out in space, Hipparcos had a clear and steady view.

OBSERVING TIPS

The head of the bull. The Hyades fit in the triangular head of Taurus, the bull. Look for it on a crisp December night to the right of Orion. With a little bit of imagination you can see the downward-angled head running into Orion's upraised shield.

Aldebaran. The eye of the bull is an orange star named Aldebaran, not to be confused with the doomed planet of Alderaan. Though not part of the Hyades cluster (it is even closer to Earth), Aldebaran is a beautiful sight on its own and enhances the cluster's beauty.

TAURUS REGION IN WINTER; 30-DEGREE FIELD OF VIEW

N

M36

M37

5

NGC 2158

Crab Nebula

Pleiades

Zeta Tauri

TAURUS

Hyades

E

S

W

ORION

MESSIER 24

BEAUTY: ★★★✦
BRAGGING RIGHTS: A beautiful sight
HOW EASY IS IT TO SEE? Best with binoculars or small telescope
BEST TIME TO SEE IT: Summer (in Sagittarius)
TYPE: Star Cloud
DISCOVERED: 1764 by Charles Messier

NOTES

When you look at the teapot of Sagittarius (page 35) you're looking toward the core of the Milky Way. Unfortunately, there are vast clouds of dust between us and the core, so we don't actually see very much.

M24 appears at first to be a rectangular patch of light in front of dark clouds. But appearances are deceiving in space—it's actually a hole in the dust clouds through which we can see deeper into the Milky Way. M24 is like a window that lets us see halfway to the center of the galaxy.

Messier 24's stars are so far away that they look flat—as if they were dots of glowing paint on the inside of a vast dome. But with a little imagination, you can look at M24 and see billowing dark clouds momentarily parting to reveal a field of stars beyond.

NGC 6603: Near the southern edge of M24, you might see a tight concentration of stars, almost like a globular cluster. This is actually the open cluster NGC 6603, but it is so far away—more than 12,000 light-years—that it appears very small. See if you can spot it.

OBSERVING TIPS

Follow the teapot. M24 is easy to find. The triangular top of the teapot points straight up at it. Wait for dark skies and a clear view of the southern horizon. If you're lucky you should be able to see it with the naked eye. Binoculars give you a wide-angle view of the whole cloud and its surroundings. But a small telescope can help you focus on the details.

A city of stars. With a small telescope you should be able to resolve the patch of light into countless stars. You're looking at another galactic arm of the Milky Way, more than 10,000 light-years away.

Dark clouds. The edges of M24 are riven with dark clouds and tendrils. There's a particularly dark pool of dust at the top center of M24. This is known as Barnard 92 and it features a lonely bright star near its center. Try to spot it.

MESSIER 22

BEAUTY: ★★★✦
BRAGGING RIGHTS: A beautiful sight
HOW EASY IS IT TO SEE? Best with binoculars or small telescope
BEST TIME TO SEE IT: Summer (in Sagittarius)
TYPE: Globular Cluster
DISCOVERED: Before 1665 by Johannes Hevelius

NOTES

Our galaxy is a flat disk of stars and dust, rotating around a spherical core. When we look up at the Milky Way in the sky, we see a band of light because we are looking at the disk edge-on. Almost all deep space objects that we see are part of the disk of the Milky Way or entirely separate galaxies millions of light-years away.

Almost, but not all. Globular clusters, like Messier 22, are an exception. Instead of orbiting along the flat disk like everyone else, globular clusters buzz around the galaxy's core like bees, orbiting in a spherical cloud often known as the galactic halo.

Globular clusters are some of the oldest features of the galaxy. They are filled with old stars and lack the clouds of primordial gas that give birth to new ones. Eventually, the stars in Messier 22 will all die. Or perhaps M22 will someday be ripped apart as it passes through the dense disk of the Milky Way.

Either way, when you look up at Messier 22, consider that it was there long before the Earth formed, and that it will still be up there, in the galactic halo, long after our own sun dies.

OBSERVING TIPS

Look toward the teapot. M22 is another showpiece object around Sagittarius. You can find it a few degrees to the left and north of the very top of the teapot. When you sweep the area with binoculars you are bound to see this compact, fuzzy globe.

It's full of stars. When it was discovered, telescopes weren't good enough to reveal more than a patchy cloud. Today, even a small telescope will show some of the thousands of stars that make up this jewel. Try medium and high magnification to get a better view.

METEOR STORM

BEAUTY: ★★★✦
BRAGGING RIGHTS: A beautiful sight
HOW EASY IS IT TO SEE? Just look up, and wait
TYPE: Special event
DISCOVERED: Known since antiquity

NOTES

Sometimes when stargazing, you need to hunch over an eyepiece, tracing the dust lanes in the Andromeda Galaxy or waiting for a brief moment of steady air to see Mars's features. At other times, however, you just need to relax on a lawn chair, look up at the night sky, and take it all in. Meteor storms are firmly in the latter camp.

Space is filled with dust and debris from comets and asteroids—but it's not evenly distributed. Instead, clumps of dust follow the same orbital paths of their origin. Dust spewed from comets leaves a trail that follows the comet's orbit. Sometimes those dust-filled orbits cross Earth's path in space.

When that happens, the particles of dust slam into our atmosphere at up to 50 miles per second. The dust turns into a shooting star, falling down from the sky in a fiery line. Most of the time, when we cross paths with a dust swarm, we see around 60 shooting stars per hour—what we call a meteor shower.

But every decade or so, we happen to hit a particularly dense patch, and then you can see shooting stars falling like rain. Spotting 60 shooting stars per minute is not uncommon, and extraordinary events feature several meteors per second. That's a meteor storm and it is not to be missed.

OBSERVING TIPS

Recurring showers. As the Earth goes around the sun, it passes through several dusty patches left behind by orbiting comets. The most famous tend to reliably result in at least a meteor shower and occasionally bloom into a full-blown storm; during a meteor shower, the meteors appear to emerge from a specific part of the sky. For this reason, they're usually associated with specific constellations (listed below).

- Quadrantids (Boötes): January 3rd to 4th.
- Eta Aquarids (Aquarius): May 4th to 5th.
- Perseids (Perseus): August 12th to 13th.
- Leonids (Leo): November 16th to 17th.
- Geminids (Gemini): December 13th to 14th.

Stay warm. The best time to see a meteor shower is usually after midnight, when the night-side of Earth is facing forward on its orbit. You'll be outside for a couple of hours, watching the show, so make sure you stay warm. Get some lawn chairs, blankets, some good friends, and just stare up at the sky.

Find the radiant. You might notice that many shooting stars seem to start at one area in the sky and radiate away in random directions. The center area is the point at which the Earth is intersecting the dust swarm. Recurring showers are usually named after the constellation they're coming from. The Leonids, for example, seem to radiate out from the constellation Leo.

DUMBBELL NEBULA (MESSIER 27)

BEAUTY: ★★★✦
BRAGGING RIGHTS: A beautiful sight
HOW EASY IS IT TO SEE? Best with small telescope
BEST TIME TO SEE IT: Summer (in Vulpecula)
TYPE: Planetary Nebula
DISCOVERED: 1764 by Charles Messier

NOTES

Deep in the core of our sun, the forces of gravity and nuclear fusion battle for supremacy. The sun's immense gravity compresses the core, driving temperatures to millions of degrees. The temperature is high enough to spark nuclear fusion: hydrogen gets converted into helium, releasing massive quantities of energy in the process. The release of energy—essentially a nuclear explosion—pushes out against gravity and keeps the sun from collapsing.

A few billion years from now, long after this book goes out of print, the sun will start to run out of hydrogen. Near the end of its life the sun will expand into a red giant as the pressures and temperatures of the core spike, and the sun's outer layers get blown into space.

Gravity will continue to compress the core, until temperatures reach over 100 million degrees. The high temperatures will ionize the outer layers of the sun, causing them to glow like neon lights.

For thousands of years, our dead star will be wrapped in a luminous shroud, visible for light-years around. Earth will be long gone by then, either swallowed up by the expanding sun or flung away into space by the violence of the collapse. The beautiful shell of glowing gas, which might look something like the Dumbbell Nebula, will be the solar system's tombstone.

And if humans have not traveled out to the stars by then, it will be ours too. Perhaps some future alien astronomer will see the glowing nebula through a telescope. If so, I hope they give it a grand name and one befitting its history.

OBSERVING TIPS

A planetary nebula. This beautiful object was the first planetary nebula discovered—so called because they are small and round, like ghostly planets. Messier 27 is easily visible with binoculars, though somewhat hard to find without a bright star to guide us.

Dumbbell or hourglass? Even at low power you can see it consists of two fuzzy blobs. I think it looks more like an hourglass than a dumb-bell. What do you see?

Devil in the detail. At high magnification you might be able to see turbulence in the bright parts. It's devilishly hard to see, but with steady skies and patience you might get there.

CYGNUS REGION IN SUMMER; 30-DEGREE FIELD OF VIEW

N

Blinking Planeta..

Epsilon Lyrae

Deneb

CYGNUS

Vega

Epsilon Cygni

LYRA

Veil Nebula

ε2

Ring Nebula

Albireo

E

W

Dumbbell Nebula

VULPECULA

A FIREBALL METEOR

BEAUTY: ★★★⭒
BRAGGING RIGHTS: A beautiful sight
HOW EASY IS IT TO SEE? Lots of luck required
TYPE: Special event
DISCOVERED: Known since antiquity

NOTES

We don't know everything that's out there. Seeing things out in space is not just about distance: it's also about size. We can see Jupiter with the naked eye, but Ceres—the largest object in the Asteroid Belt—is invisible without a telescope, even though it is much closer than Jupiter.

Most of the time, this doesn't matter. Life on Earth does not significantly change when a new asteroid is discovered. But someday—maybe tomorrow, maybe 10,000 years from now—we will discover a new asteroid on a collision course with Earth, and then we will want as much warning time as possible.

Stuff hits Earth all the time. Shooting stars (see page 136) are little pebbles blazing through the atmosphere spectacularly, but harmlessly. Larger objects can have a more serious effect, however.

Boulder-size objects sometimes hit our planet, lighting up the night sky. When they're brighter than Venus, they're known as fireball meteors, and their unexpected appearance can jolt you with surprise.

In 2013, a 12,000-ton object, probably 60 feet across, slammed into the skies over Russia. Thousands saw it streak across the daytime sky, brighter than the sun. The object exploded with 20 times the energy of the first atomic bomb. It exploded before it hit the ground, injuring many with its shockwave.

And no one saw it coming. Robotic telescopes are currently scanning the sky, looking for other interlopers, and every year we discover more objects likely to hit Earth. But we'll never find them all. Some objects are just too small, so we will always have the thrill of seeing a bright fireball meteor, unexpected and unannounced, appear amid the calm and steady stars. And the possibility of a more sinister visit remains as well.

OBSERVING TIPS

Keep looking up. Until we can discover and track tiny objects in space, we will not be able to predict when a fireball meteor will appear. Your best chance of seeing one is to keep looking up. Meteor showers are the best time, obviously, but anytime you look at the night sky you have a chance of seeing a fireball.

Take note of details. You won't have time to take a picture of it, so memory is all you'll have. Describe what you saw as soon as you can. Did you see colors? Did you see sparks? Did you hear a sound? Putting your experience into words is the best way to remember it.

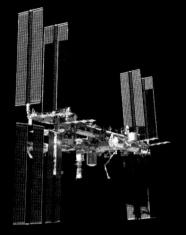

"I've seen it!"

THE INTERNATIONAL SPACE STATION

BEAUTY: ★★★
BRAGGING RIGHTS: You saw a space mission in progress
HOW EASY IS IT TO SEE? Just look up at the right time
TYPE: Orbiting space station
DISCOVERED: 1998

NOTES

No one knows the name of the first person to land on the Hawaiian islands. Was it a settler from the Marquesas? Was it a navigator-priest from Tahiti? No one knows. But the voyage was as audacious as any other tale of human exploration. Imagine setting out into the endless ocean, without maps, without GPS, and without engines. A dozen people on a rugged wooden boat with a flimsy sail were enough to explore an island 2,000 miles away.

At any given time, the International Space Station is no more than 270 miles away. Many of us have driven longer than that without stopping. But instead of a tropical paradise, the settlers of space must deal with a hard vacuum, solar radiation, and micrometeorite collisions—all while traveling at 17,000 miles per hour.

Space is the most challenging environment we've ever explored. The combined resources of a dozen countries are barely enough to keep six humans alive in orbit.

And yet they're there. Ever since 2000, humans have had a permanently inhabited colony in space. Men and women have been working and living in space for almost 20 years. And almost none of us know their names.

OBSERVING TIPS

Sighting opportunities. NASA has an excellent website to help you determine when the International Space Station is visible near you: http://spotthestation.nasa.gov/sightings/

Time in the sun. The sun has to be shining on the ISS for you to see it. But if the sun is visible for you, then its glare will make it impossible to see the Station. How is it possible to see ISS then?

Have you ever seen a brightly lit cloud, high in the sky after sunset? The sun has set for you, but since the cloud is higher up, it can still see the sun over the horizon—this is one simple proof the Earth is round. The ISS is 20 times higher than most clouds, so it can be in sunlight even though you're in darkness below.

Don't wait too long. Space is a harsh environment. Equipment degrades; orbits decay. Eventually, it will cost too much to keep the ISS in orbit, and after the last crew comes home, it will be instructed to fall from the sky and plunge into an ocean. As of this writing, Space Station operations are funded through 2024, so try to see it before then!

ALGOL

BEAUTY: ★★☽
BRAGGING RIGHTS: You saw the Demon Star
HOW EASY IS IT TO SEE? Best with binoculars or small telescope
BEST TIME TO SEE IT: Winter (in Perseus)
TYPE: Eclipsing Binary
DISCOVERED: Known since antiquity

NOTES

In the movie *Batman Begins*, Ra's al Ghul starts out as a mentor to Bruce Wayne, but he is eventually unmasked as a super villain. The star known to Arab astrologers as Ra's al Ghul—"Algol" to us Westerners—was similarly mysterious and required much effort to uncover its true nature.

In the seventeenth century, astronomers noticed that al Ghul's light dimmed every 69 hours and stayed dim for 10 hours. After that, it regained its original brightness as if nothing had happened. What could be going on?

In the eighteenth century, amateur astronomer John Goodricke proposed that maybe Algol was dimming because a dark body was passing in front of it. It was a plausible theory, but no telescope then could magnify sufficiently to confirm the hypothesis.

In the nineteenth century, spectrographs were used to measure motion in the stars, and Ra's al Ghul was revealed to be moving back and forth with a period of 69 hours or so. Most likely a dim star was gravitationally tugging on Algol as it spun around, and as it passed in front of its brighter sibling, it partially eclipsed it.

Finally, in the twenty-first century, an ultra-high-resolution interferometer took pictures of Algol and revealed its hidden companion. The mystery of Ra's al Ghul was conclusively solved. Batman would be proud.

OBSERVING TIPS

The Ghoul's Head in Perseus. Look for Algol in autumn, in the constellation Perseus. It usually shines at magnitude 2.1—about as bright as any star in the Big Dipper. But once every 69 hours it fades to magnitude 3.3—more than three times dimmer.

Compare and Contrast. At its brightest, Algol will be as bright as Gamma Andromedae (Almach), which is nearby. When eclipsed, it will be as faint as Epsilon Persei. Track your perception of its brightness and see how close you can come to determining how long it takes to vary.

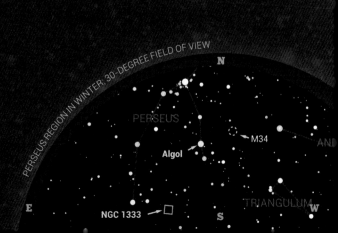

BETELGEUSE

BEAUTY: ★★✔
BRAGGING RIGHTS: You saw a red giant
HOW EASY IS IT TO SEE? Just look up
BEST TIME TO SEE IT: Winter (in Orion)
TYPE: Star
DISCOVERED: Known since antiquity

NOTES

Betelgeuse is big—really big. But space is bigger—unimaginably bigger. Betelgeuse is a supergiant star, almost a thousand times bigger (in diameter) than our own tiny sun. It is one of the largest stars we know about and the tenth-brightest star in our sky (eleventh if you count the sun).

If Betelgeuse were at the distance of Pluto, it would appear 24 times larger than the full moon. It would be an amazing sight—if any humans survived its massive energy output.

But space is big. Really big. Betelgeuse is almost 650 light-years away—about 40 million times farther away from us than the sun. At that distance,

Betelgeuse looks tiny. To get any idea of just how big space is, imagine if you could see a bee from a mile away. Betelgeuse appears 30 times smaller than that.

And yet, in the 1920s, scientists were able to magnify Betelgeuse enough to measure its diameter. It is one of the few stars large enough that we can directly measure its size. Space is big, but human ingenuity is up to the challenge.

OBSERVING TIPS

On the shoulder of Orion. Betelgeuse is easily seen in winter. It is the bright reddish star forming Orion's left shoulder (as seen from us).

Red and blue. Stars are normally too faint to trigger our color receptors. But Betelgeuse is bright enough to appear noticeably colorful. Compare red Betelgeuse to blue-white Rigel—the star on the bottom right of Orion.

Variable star. Like other red supergiants, Betelgeuse is slightly variable: it expands and contracts as the forces of gravity and nuclear fusion fight for dominance. Sometimes Betelgeuse can be as faint as magnitude 1.3—fainter than Rigel. Other times it can be as bright as Vega.

Supernova? In 100,000 years or so, Betelgeuse will run out of nuclear fuel and gravity will win once and for all. The outer layers of the star will fall down to the solid core, accelerating all the way. When they hit, the resulting explosion will be a supernova.
For months Betelgeuse could be as bright as the moon.

ORION REGION IN WINTER; 30-DEGREE FIELD OF VIEW

N

Betelgeuse

ORION

MONOCEROS

M78

Zeta Orionis

Orion Nebula

Rigel

E

S

W

ERIDANUS

SIRIUS

BEAUTY: ★★⭒
BRAGGING RIGHTS: You saw the brightest star in the night sky
HOW EASY IS IT TO SEE? Just look up
BEST TIME TO SEE IT: Winter (in Canis Major)
TYPE: Star
DISCOVERED: Known since antiquity

NOTES

Humans love records: the fastest airplane, the deepest ocean, the wealthiest movie star. But to have a record, you first need to measure. Sirius is the brightest star in the night sky—a fact known since antiquity—but how bright is it? What units do we use for brightness?

Two thousand years ago, Claudius Ptolemy looked up from Alexandria, Egypt, and created a catalog of stars (cataloging is another favorite human pastime). He assigned each star one of six magnitudes, according to their brightness (as he perceived them). Stars of the first magnitude were the brightest, while stars of the sixth were at the faintest limit of visibility.

After the Renaissance, catalog makers followed Ptolemy's convention but subdivided magnitudes with decimal points. A 2.1 magnitude star was slightly fainter than 2.0. Magnitudes ceased to be discrete categories and became continuous values, like distance. In the 1800s, astronomers formalized the scale by defining magnitude 6 to be 100 times fainter than magnitude 1. That means each magnitude is about 2.5 times fainter than the previous one. They chose the star Vega to be at exactly magnitude 0. A star 2.5 times fainter than Vega was magnitude 1, and one 2.5 times brighter was magnitude -1.

In the new system, Sirius turned out to be magnitude -1.46, and no other star in our sky (save the sun) is brighter.

OBSERVING TIPS

The Dog Star. Sirius is the brightest star in the constellation Canis Major ("Big Dog," in Latin) and in Roman times, the star itself was known as the Dog Star. To find it, start at the three stars in Orion's belt and follow their direction down and to the left; they will point straight at Sirius, the brightest star in the sky.

A remarkable number of cultures associate Sirius with dogs, from the Chinese ("Heavenly Wolf") to Alaskan Natives ("Moon Dog").

Hidden companion. Sirius has a white dwarf companion, known as Sirius B, but it's sometimes called "The Pup." The Pup takes 50 years to orbit around its parent, and it will be at its greatest distance from Sirius A around the year 2025. With a good-size telescope and high magnification you might be able to spot this diminutive companion.

CANIS MAJOR REGION IN WINTER; 30-DEGREE FIELD OF VIEW

ALBIREO

BEAUTY: ★★★
BRAGGING RIGHTS: A beautiful sight
HOW EASY IS IT TO SEE? Best with binoculars or small telescope
BEST TIME TO SEE IT: Summer (in Cygnus)
TYPE: Star
DISCOVERED: Known since antiquity

NOTES

I have trouble using faucets sometimes. A red mark on a tap is supposed to indicate hot water, while a blue mark signifies cold water. But that's backwards for an amateur astronomer!

Light, whether reflected by a faucet sign or emitted by a star, comes in different frequencies or pitches—much like sound, and our eyes and brain interpret different pitches as different colors. High-frequency light appears blue to us, while low-frequency light is red. It turns out that hotter things emit more high-frequency light. When it comes to stars, blue is hotter than red.

We think of red as hot because we think of glowing hot coals. And it's true that red stars are hot—I wouldn't want to touch one with my bare hands. A red dwarf shines at 3,000 °F—about the same temperature as hot coals. But blue stars are even hotter: a blue giant's surface is more than 30,000 °F, hot enough to melt any material on Earth.

Albireo is a beautiful demonstration of this difference; it is a double-star in which one member is reddish while the other is blue. Albireo A, the brighter member, is an amber-colored giant with a temperature of 4,000° F, while its companion, Albireo B, is a smaller blue star shining at more than 13,000° F.

As you look at this pair, try to ignore everything you learned from faucets, and remember that blue is hotter than red.

OBSERVING TIPS

The base of the Northern Cross. Look up in the sky on a summer night and you'll see three bright stars forming the Summer Triangle: Deneb, Vega, and Altair. Deneb forms the head of the Northern Cross—a group of five stars shaped like a Christian cross. Albireo is the star that forms the base of the cross.

Beautiful contrast. Albireo is probably the finest double star in the sky because of its contrasting colors. A telescope at low power (maybe 30×) is best to show the contrast. Try unfocusing the stars a little bit to make the colors more obvious.

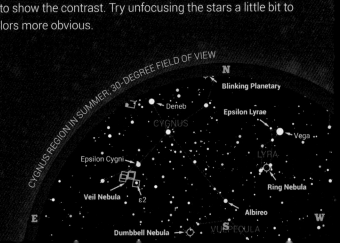

CYGNUS REGION IN SUMMER; 30-DEGREE FIELD OF VIEW

N

Blinking Planetary

Deneb

Epsilon Lyrae

CYGNUS

Vega

LYRA

Epsilon Cygni

Ring Nebula

Veil Nebula   ε2

Albireo

E

W

S

VULPECULA

Dumbbell Nebula

DOUBLE CLUSTER (CALDWELL 14)

BEAUTY: ★★★
BRAGGING RIGHTS: A beautiful sight
HOW EASY IS IT TO SEE? Best with a small telescope
BEST TIME TO SEE IT: Winter (in Perseus)
TYPE: Open Cluster
DISCOVERED: Known since antiquity

NOTES

Astronomers—amateur and professional alike—love catalogs. If you want to make sense of our universe, or even if you just want to see the sights, you need to know what's out there and where to look.

Charles Messier's catalog of deep-sky objects is justly famous, but as I point out on page 158, he never meant to create a listing of the best sights out there. Patrick Caldwell-Moore, on the other hand, did.

To call Sir Patrick Moore an amateur astronomer is unforgivably stingy. He was the ultimate amateur astronomer: author of more than 100 books, host of the BBC's *The Sky at Night* program, and creator of the Caldwell Catalog of deep-sky objects.

Patrick Moore created his catalog in 1995 to supplement Messier's. He wrote down the best nebulae, galaxies, and clusters omitted by Messier ("missed" is too strong a word). Moore then mailed off the list to *Sky and Telescope* (today he would have just posted it on the Internet) and it's been famous since.

The Double Cluster in Perseus is a perfect example of the kinds of objects in the Caldwell Catalog. It's been known since the ancient Greeks and it is a favorite target for amateurs. Now it is finally listed in an accessible catalog. Thanks to Patrick Moore.

OBSERVING TIPS

A fuzzy cloud. On a dark autumn night you might be able to see the Double Cluster with the naked eye as a fuzzy star between Cassiopeia and Perseus. Binoculars will quickly reveal its dual nature, though you may not be able to resolve all its stars. But point a telescope at it and you'll see two tight groups of brilliant stars.

Use low power. The two clusters are 25 arcminutes apart—almost the width of the full moon. Use a telescope with low power to see as much of the two clusters as possible. Excessive magnification lessens the view considerably.

Compare against the Seven Sisters. The Pleiades (page 76) are much closer to us, but the Double Cluster has more stars. Both should be visible on the same night, so take a moment to compare the two.

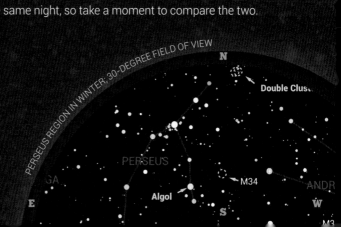

PERSEUS REGION IN WINTER, 30-DEGREE FIELD OF VIEW

N

Double Clus.

PERSEUS

M34

VGA

ANDR

Algol

E

S

W

M3

EPSILON LYRAE

BEAUTY: ★★★
BRAGGING RIGHTS: A beautiful sight
HOW EASY IS IT TO SEE? Best with a small telescope
BEST TIME TO SEE IT: Summer (in Lyra)
TYPE: Star
DISCOVERED: 1779 by William Herschel

NOTES

In *A New Hope* (I'm old enough to remember when it was just called *Star Wars*) Luke Skywalker looks at the horizon to see two suns setting. It is one of the many iconic scenes from the movie, and in reality, it's a common sight. Many sun-like stars in our galaxy are part of a two-star system. Less common are double-double systems like Epsilon Lyrae.

If you look at Epsilon Lyrae with binoculars, you'll notice that it's actually two stars, separated by less than 3 arcminutes (about 3 Jupiter diameters, as seen from Earth). One star is known as ε1 (epsilon 1) while its partner is ε2.

But if you look with a telescope at medium magnification (×150) you'll see that both stars are themselves double stars. A double-double star! If Tatooine were orbiting one of the suns of this system, Luke would see four suns.

If Luke were on a planet orbiting one of the stars of ε1, he'd see one star as big as the sun. The second star of ε1 would be 120 AUs away— three to four times farther away than Pluto. It would be much fainter than the sun, but still 50 times brighter than the full moon. The pair of stars in ε2 would be much fainter still. They are 1/6th of a light-year away, about 50 times closer than Sirius. They would appear as very bright stars, of magnitude -8 or -9, fainter than the full moon, but far brighter than any star in our sky.

OBSERVING TIPS

Summer Skies. Look for Epsilon Lyrae during summer months when Vega—the fifth brightest star—is high in the sky. Vega forms a small triangle with two other significantly fainter stars. One of those fainter stars is Epsilon Lyrae.

Split the Double. Good binoculars and steady skies are enough to split Epsilon Lyrae into a pair. With excellent skies—and sharp eyesight—you might be able to split the pair with the naked eye. Even if you can't see it, you should still be able to tell that the star is slightly elongated.

The Double-Double. Use a telescope to split each of the components. If you can't split them at first, try increasing the magnification. If you still can't see all four stars, you may need to wait until a steadier night.

CYGNUS REGION IN SUMMER, 30-DEGREE FIELD OF VIEW

N

Blinking Planetary

Deneb

Epsilon Lyrae

CYGNUS

Vega

Epsilon Cygni

LYRA

Veil Nebula

ε2

Ring Nebula

Albireo

E

S

W

Dumbbell Nebula

VULPECULA

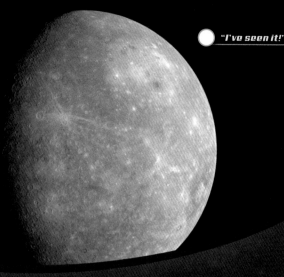

MERCURY

BEAUTY: ★★★
BRAGGING RIGHTS: You saw the closest planet to the sun
HOW EASY IS IT TO SEE? Best with a small telescope
TYPE: Planet
DISCOVERED: Known since antiquity

NOTES

From Earth we see the solar system edge-on. Looking toward the sun we see Mercury moving back and forth on a nearly straight line across the sun. At one extreme on the line, Mercury rises in the morning before the sun does. At the other extreme, it sets in the evening just after the sun. The ancient Greeks didn't realize the two sightings were of the same planet. They called the evening star Hermes (Mercury) and the morning star Apollo.

In the nineteenth century, astronomers once again thought Mercury might have a double. Armed with Isaac Newton's rigorous equations for planetary motion, they tried to predict Mercury's position. They failed: no matter how carefully they checked their numbers, Mercury was always slightly off. Scientists theorized that an unknown planet was pulling

Mercury off course. They called this hypothetical planet Vulcan. For decades, astronomers searched for Vulcan. Several claimed to have found it, but no observation ever amounted to much. Finally, in 1915, the truth was discovered: Mercury's orbit was off, not because a planet was pulling it, but because Newton's theory was wrong! Einstein's theory of space-time superseded Newton's and perfectly predicted Mercury's orbit—Vulcan was no longer needed, and Mercury was alone once more.

OBSERVING TIPS

In the glare of the sun. Mercury is the most challenging of the five classical planets to observe. It never strays far from the sun and is generally only visible right after sunset or just before dawn. Part of the thrill of seeing this tiny planet—the smallest of the eight, now that Pluto has been demoted—is in racing to observe it before sunrise overwhelms it, or before it follows the sun below the horizon.

Wait for maximum elongation. The time when Mercury is farthest from the sun (as seen from Earth) is called maximum elongation. These are the best times to see Mercury away from the sun's glare.

See a transit. Mercury passes directly in front of the sun once a decade or so. With proper equipment, you can see a tiny black disk move swiftly across the sun's face.

BEST DATE IN THE MORNING TO SEE MERCURY	BEST DATE IN THE EVENING AFTER SUNSET
May 18, 2017	July 30, 2017
September 12, 2017	November 24, 2017
January 2, 2018	March 16, 2018
April 30, 2018	July 12, 2018
August 27, 2018	November 7, 2018
December 15, 2018	February 27, 2019
April 12, 2019	June 24, 2019
August 10, 2019	October 20, 2019
November 28, 2019	February 11, 2020
March 24, 2020	June 5, 2020
July 23, 2020	October 2, 2020

RING NEBULA (MESSIER 57)

BEAUTY: ★★★
BRAGGING RIGHTS: A beautiful sight
HOW EASY IS IT TO SEE? Best with a small telescope
BEST TIME TO SEE IT: Summer (in Lyra)
TYPE: Planetary Nebula
DISCOVERED: 1779 by Antoine Darquier de Pellepoix

NOTES

The Ring Nebula is one of the first sights I ever saw with my telescope. Using my well-thumbed copy of *The Field Guide to the Stars and Planets*, I pointed my telescope at Vega and then moved from star to star. Soon, a beautiful, ghostly smoke ring drifted into view—right where it was supposed to be. The sky charts were like a key to a secret world.

Without catalogs and charts to guide us we would never find the wonders of the night sky. But who created the catalogs? How did they find all those beautiful objects in the sky? The most famous list of interesting objects is the Messier Catalog, compiled by Frenchman Charles Messier in the late 1700s. Ironically, he wasn't interested in the objects at all—his goal was to make a list of objects to avoid.

Charles Messier was obsessed with comets—those awesome but unpredictable visitors in the sky. His tireless work to discover new comets earned him a job as Astronomer of the Navy and membership in the Royal Academy of Sciences in Paris. One day he stumbled on a faint glow in Taurus that looked like a comet. But it wasn't. It never moved from its position. To avoid being fooled in the future, Messier began a catalog of fuzzy, comet-like objects. Messier 1, the Crab Nebula, was the first object. By the time he got to Messier 57, like-minded astronomers were sending him their own non-comet discoveries. Today, Messier's list consists of 110 deep-sky objects, all within reach of amateur instruments, and they are well worth the effort to track down.

OBSERVING TIPS

Smoke ring in Lyra. Look up in the sky in summer and you should see Vega—one of the brightest stars in the sky. Vega is part of a small triangle with two fainter stars. Attached to the triangle at one point is a parallelogram. Look at the short side of the parallelogram opposite Vega to find the Ring Nebula.

Magnify. Messier 57 is small and relatively bright. It should respond well to moderate magnification. Nevertheless, you might have to use averted vision to see more detail. You'll notice that the ring is well-defined and distinctly oval. See if you can notice any detail. Are parts of the ring brighter than others?

Shroud of a Dead Star. Like the Dumbbell Nebula (page 138) Messier 57 is a planetary nebula—the cast-off shell of a red giant after it collapsed and turned into a white dwarf. It helps to think of the Ring Nebula, not as a ring, but as a spherical shell.

CYGNUS REGION IN SUMMER; 30-DEGREE FIELD OF VIEW

N

Blinking Planetary

Deneb

Epsilon Lyrae

CYGNUS

Vega

LYRA

Epsilon Cygni

Ring Nebula

Veil Nebula

ε2

Albireo

E

S

W

VULPECULA

Dumbbell Nebula

TITAN

BEAUTY: ★★★
BRAGGING RIGHTS: You saw a moon of Saturn
HOW EASY IS IT TO SEE? Best with a small telescope
TYPE: Moon
DISCOVERED: 1655 by Christiaan Huygens

NOTES

"It is a riddle, wrapped in a mystery, inside an enigma." Churchill was referring to Russia in 1939, but he could have also been talking about Titan. Titan was discovered by Dutch physicist Christiaan Huygens in 1655. He saw a bright dot next to Saturn, calculated its orbit and rough mass, and named it Titan. For almost 300 years, nothing more was learned about it.

In 1944, while Churchill was busy with a world war, another Dutchman, Gerard Kuiper, detected methane on Titan. Methane? But how? Moons are too small to hold on to an atmosphere. But Kuiper realized that the sub-zero temperatures on Titan would keep the methane from evaporating away into space. Unfortunately, that thick atmosphere obscured all surface features. In 1980 the Voyager 1 space probe—

launched on a descendant of a World War II weapon—the V-2—flew close to Titan but was unable to unravel its mysteries. Photos showed nothing but an orange haze.

In 2004, the Cassini spacecraft flew by Titan, this time armed with another World War II invention: radar. Piercing through Titan's clouds, it revealed a fantastical terrain with hydrocarbon oceans and wispy methane clouds. Finally, a probe, appropriately named after Huygens, landed on Titan and sent back color images of its rock-strewn surface.

But mysteries remain. With hydrocarbon seas and a nitrogen-methane atmosphere, Titan resembles a primordial Earth—albeit at a much colder temperature. Could the seeds of life be found among its tidal pools and shores? No one knows. Titan is still an enigma.

OBSERVING TIPS

Ring-plane orbit. Titan orbits on the same plane as Saturn's famous rings, but much farther out from the planet. Imagine if the rings could extend 10 times farther out; Titan would be orbiting at the edge. Try to mentally expand the rings to estimate Titan's orbit.

Wait for it to move. Titan takes about 16 days to orbit the planet. That means in a little over a week it will be on the opposite side of Saturn. If you find a bright star near Saturn, record its position on paper. Check again a few days later and see if it has moved. With enough patience you may be able to trace out the whole orbit.

Tholin haze. Carl Sagan came up with the term tholin to describe the kinds of chemicals in Titan's atmosphere that give it its orange color. Can you see a faint tint of orange on Titan with your own eyes?

URANUS

BEAUTY: ★★★
BRAGGING RIGHTS: You saw the 7th planet
HOW EASY IS IT TO SEE? Best with a small telescope
TYPE: Planet
DISCOVERED: 1781 by William Herschel

NOTES

As of this writing, fewer than 50 people have been president of the United States, and only 12 men have ever walked on the moon. But as exclusive as those clubs are, William Herschel belongs to an even more exclusive club: he is one of only two people to have discovered a planet in our solar system. William Herschel was an amateur astronomer (he made money as a music tutor) in an age when amateurs could make significant contributions. Herschel built his own telescopes in his spare time and regularly exceeded both the size and quality of what professionals were using.

Before astrophotography, automated telescopes, and computerized catalogs, the only way to discover something was to look at the sky night after night, month after month. Herschel had begun a personal project to catalog double stars. He scanned the sky methodically, inspecting every bright star to see if it had a companion. In 1781, he wrote a letter to London's Royal Society listing 269 double stars, which claimed the North Star was a double star. Astronomers were dubious: what are the chances that a no-name amateur with homemade gear could detect a companion to one of the most famous and well-observed stars in the sky? And yet, a few months later, his discovery was confirmed, and Herschel received due fame. That same year, Herschel spotted an unusual star in Taurus. In his telescope, it appeared round and a little fuzzy. For a long time, Herschel thought he had discovered a comet, but other astronomers soon computed its orbit and realized that it moved in a stately circle, far beyond the orbit of Saturn: it was a new planet. Herschel proposed naming the planet Georgium Sidus, in honor of King George III. But other astronomers prevailed and called the planet Uranus. Nevertheless, King George was impressed, and he gave Herschel an annual stipend of £200 (about $50,000 today) to spend watching the stars.

OBSERVING TIPS

Can you find it? Uranus shines at magnitude 5.3, faint to be sure, but easily visible to the naked eye if you have dark skies. Unfortunately, it gets lost among all the other faint anonymous stars in the sky, but if you know where to look, you should be able to find it. Consult a good star chart for Uranus's position and see if you can spot it.

A blue-green disk. The telltale sign that you're looking at a planet is if you can see it as a small, round disk instead of a point of light. You'll need a telescope at high magnification to see it. Look directly at it. Can you see any color? With a little patience you should see a pale blue-green hue.

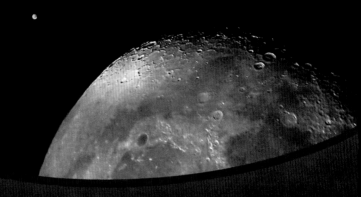

PLANETARY CONJUNCTIONS AND OCCULTATIONS

BEAUTY: ★★★
BRAGGING RIGHTS: A beautiful sight
HOW EASY IS IT TO SEE? Best with small telescope
TYPE: Special event
DISCOVERED: Known since antiquity

NOTES

We have an amazing capacity for pattern recognition. The earliest peoples mastered the recurring cycle of the moon to keep time. The Egyptians used the appearance of Sirius in the morning twilight (around June 22nd at that time) as a sign that the waters of the Nile would begin to rise (generally around June 25th). The sky was full of patterns that affected daily life.

But sometimes our pattern recognition goes awry. To the ancients, the sight of two bright planets close together in the sky must have begged for explanation: What did it portend? Good? Ill? No matter what happened next, the death of a king or the end of a drought, people would attribute it to the planetary alignment. Astrology is pattern recognition devoid of scientific rigor. It is pareidolia: seeing the face of Elvis in burnt toast. And yet the beauty of these patterns cannot be denied. Seeing Jupiter near Venus or Mars near the moon is a special treat. Alone in the sky, any planet is lost amid the myriad anonymous stars, but it's impossible to ignore the sight of two or more bright planets together. The familiar sky, taken for granted in our busy lives, again becomes a source of wonder. All because of a pattern.

OBSERVING TIPS

Check your local listings. Conjunctions and occultations don't come on a recurring schedule. Check the calendar on publications such as *Astronomy* and *Sky & Telescope* for upcoming events. Conjunctions between the moon and a planet are the most common and occur when the moon and a planet appear in the sky together. In rare cases they get so close that the moon covers up the planet: this is an occultation, not to be missed. Seeing two planets together can be thrilling, but seeing three near each other—often called a trio—is a rare but rewarding treat.

Close up views. In some cases two planets get close enough that you can see them together in a telescope. These views are fantastic because you can see them as real objects right next to each other. The planets need to be within half a degree for this to be possible—roughly a full moon in diameter. But the closer they are, the more you'll be able to magnify.

Other conjunctions. Conjunctions between a planet and a deep-sky object are also wonderful sights through a telescope. Venus and other planets sometimes appear near the Pleiades (page 76).

VIRGO CLUSTER OF GALAXIES

BEAUTY: ★★★
BRAGGING RIGHTS: You saw a galaxy cluster
HOW EASY IS IT TO SEE? Telescope required
BEST TIME TO SEE IT: Spring (in Virgo)
TYPE: Galaxy Cluster
DISCOVERED: 1931 by Edwin Hubble & Milton Humason

NOTES

The simple quest to determine the distance to the stars led to one of the most profound discoveries in history. How far away is that wispy cloud in Andromeda? Is it a glowing cloud of gas inside our galaxy? Or is it something much larger, and more distant? Henrietta Leavitt (page 176) provided the answer: the Andromeda Galaxy is a vast collection of stars millions of light-years away from us. The universe suddenly multiplied in size.

But the shocks were only beginning. Edwin Hubble and his colleagues used Leavitt's technique to determine the distance to hundreds of galaxies—including those in the Virgo Cluster of Galaxies. All the distances were in millions of light-years, and they allowed Hubble to get a sense for our galaxy's place in the universe.

But Hubble also noticed something curious. Almost every galaxy he measured had a red shift in its spectrum: it was the visual equivalent of a train whistle changing pitch as it churned away from us. It meant that all these galaxies were moving away from us. Even more astonishing, the speed at which they were receding depended on their distance. Farther galaxies were moving faster! In fact, the relationship is so consistent that you can guess a galaxy's distance by knowing how fast it's moving. This relationship is now known as "Hubble's Law." But what does it mean? Why are galaxies moving away from us? Our best explanation today—ridiculed at the time, but reenforced with each discovery—is the Big Bang: all matter was once consolidated at a single point before a tremendous explosion caused it to spread across the universe. Along the way, galaxies formed, and many of the galaxies we see appear to be expanding away from us.

OBSERVING TIPS

Widefield view. The Virgo Cluster is a group of galaxies—faint, but fascinating—spread out over a large region of sky. To see it, look between the constellations Virgo and Leo. Scan the line from Denebola (in Leo) to Epsilon Virginis to take in the heart of the cluster. Don't expect too much, however—most of these galaxies are faint blurs in binoculars.

Highlights. The brightest members are **M49, M87, M60,** and **M86,** all of which are elliptical galaxies. Beautiful spiral galaxies can also be found, but they are fainter. Look at **M100, M61, M88,** and **M99.**

COMA BERENICES REGION IN SPRING; 30-DEGREE FIELD OF VIEW

N

NGC 4565

OMA BERENICES

M64

Coma Star Cluster

Alpha

Virgo Galaxy Cluster

RIGA

Denebola

ANDROMEDA

M65 & M66

M87

Epsilon Virginis

20 Virginis

Rho Virginis

E

S

TRIANGULUM

W

WHIRLPOOL GALAXY (MESSIER 51)

BEAUTY: ★★★
BRAGGING RIGHTS: A beautiful sight
HOW EASY IS IT TO SEE? Telescope required
BEST TIME TO SEE IT: Spring (in Canes Venatici)
TYPE: Galaxy
DISCOVERED: 1773 by Charles Messier

NOTES

Imagine a TV series like *Downton Abbey* featuring not the fictional Earl of Grantham, but the real Earl of Rosse: William Parsons. Lord Rosse was an amateur astronomer who had the means to build successively larger and larger telescopes. His ultimate creation was the Leviathan of Parsonstown: a 50-foot long, 6-foot diameter reflecting telescope housed in a 3-story castle. It was the largest telescope in the world for more than 70 years. Such a TV series would follow Lord Rosse in his attempt to uncover the true nature of nebulae. Were they gigantic gas clouds in the process of forming new star systems? Or were they—as Lord Rosse believed—vast stellar clusters, so far away that individual stars could not be seen? Would Lord Rosse be able to prove his theories to the skeptical astronomical community? Or would his quaint ideas ostracize him from his scientific peers? With his Leviathan, Rosse stared at Messier 51, a nebula near the Big Dipper, and discovered its spiral form. His drawing of the "Whirlpool" closely matches modern astrophotos. But his painstaking studies were not able to resolve the fundamental question. Many years later, using instruments unknown to the nineteenth century, astronomers discovered that Lord Rosse was both right and wrong: some nebulae, like the Orion Nebula, are luminous gas clouds in the process of forming new stars. But others, including the spiral Messier 51, are indeed unimaginably distant clusters of stars. These "island universes" were far more numerous than Lord Rosse ever imagined, and their study ultimately revealed how the universe came to be.

OBSERVING TIPS

At the Handle of the Dipper. Start your search for M51 at Eta Ursa Majoris, the star on the end of the Big Dipper's handle. Move about 3 degrees in a line toward Beta Canum Venaticorum (see chart) and you should run into it.

Double Core. With a small telescope pointed at M51, you might notice two distinct fuzzy objects. The larger object is the core of M51, while the smaller one is a companion galaxy, known as NGC 5195. The latter is orbiting around M51 and was probably once a spiral galaxy too, but its close encounter with the larger M51 has disrupted it.

Clear Skies and Lots of Aperture. Don't expect to see spiral arms unless you have a large telescope (12" and greater) and very clear skies. A star party or a local college observatory is your best bet to see this faint galaxy in all its glory.

URSA MAJOR REGION IN SPRING, 30-DEGREE FIELD OF VIEW

N

M101

Alcor & Mizar

M109

URSA MAJOR

Eta
Ursa Majoris

M106

M51

M94

M63

Beta
Canum Venaticorum

CANES
VENATICI

Alpha

E

S

W

CALDWELL 38 (NGC 4565)

BEAUTY: ★★★
BRAGGING RIGHTS: A beautiful sight
HOW EASY IS IT TO SEE? Telescope required
BEST TIME TO SEE IT: Spring (in Coma Berenices)
TYPE: Galaxy
DISCOVERED: 1785 by William Herschel

NOTES

Compare photos of NGC 4565 with M51 (page 168). Would it ever occur to you that both are the same kind of object, just seen from a different perspective? It took many years and hundreds of observations for astronomers to agree that this was the case.

Today we know that spiral galaxies like NGC 4565 and M51 are flattened disks of stars and dust with a bulging central core. Our own Milky Way is like that too. From above it looks like a beautiful spiral. But from the side, it looks just like NGC 4565.

Imagine zooming in to NGC 4565. You'd see the galaxy grow, its faint glow expanding proportionally. Soon it would grow beyond your field of view. You'd have to move your head around to see the whole thing. You'd probably see lots of detail: faint embedded nebulae, and lots of little star clusters. Eventually, the galaxy would get so big that it would arc from one horizon to the other. You'd see a band of light splitting the sky in two. Exactly like the Milky Way.

OBSERVING TIPS

Not a Messier object. NGC 4565 is one of the few bright galaxies that Charles Messier omitted when he put together his eponymous catalog.

To find it, start at the Coma Star Cluster (page 192) and move due east a few degrees. Though brighter than some Messier objects, it's still hard to find in any but the darkest skies. NGC 4565 is probably too faint for binoculars, but even a small telescope will show something interesting.

Central bulge. Start with low power under dark skies to see NGC 4565. The first thing you'll notice is the central bulge. This is the core of the galaxy and it will appear slightly oval. Can you see the needle-like shapes extending out from the oval? That's the disk seen edge-on. Use averted vision to extend the size of the galaxy.

Dust lane. Through larger telescopes you might be able to see a thin dark lane of dusk splitting the needle lengthwise. With a smaller scope, try magnifying the central core and see if you can spot the thin band.

COMA BERENICES REGION IN SPRING, 30-DEGREE FIELD OF VIEW

N

Gamma

NGC 4565 →

COMA BERENICES

M64 →

Alpha

Coma Star Cluster

Virgo Galaxy Cluster

AURIGA

Denebola

ANDROMED.

M65 & M66

E

Epsilon
Virginis

M87

20 Virginis

S

W

Rho Virginis

POLARIS

BEAUTY: ★★↗
BRAGGING RIGHTS: You saw the North Star
HOW EASY IS IT TO SEE? Just look up
BEST TIME TO SEE IT: Year-round
TYPE: Star
DISCOVERED: Known since antiquity

NOTES

If ever you need a reminder that life is short, just stare up at Stella Polaris: the Pole Star. Astronomy is filled with cycles: the daily rotation of the world, the monthly cycle of the moon, and our annual journey around the sun. Those cycles are all human-scaled, but there are many other cycles that transcend a single human lifespan. And just as fruit flies (average lifespan, 30 days) have no concept of the seasons, we humans (average lifespan, 75 years) are blissfully unaware of the changing of the Pole Star.

Once a day, our world spins on its axis, and that axis happens to point in the direction of Polaris. When you take a long-exposure picture of Polaris, you can see star trails rotating around it. This is an illusion, of course, as it is we on Earth who are rotating.

But the axis also moves! Just as a spinning top sometimes wobbles, the Earth's axis traces a lazy circle around the sky. Polaris wasn't always the Pole Star, and some day in the future, it won't be anymore. The axis makes one complete circle every 26,000 years: about 350 human lifespans. That's the equivalent of a fruit fly contemplating a 30-year mortgage.

About 26,000 years ago humans were bravely enduring the worst of the most recent ice age. I don't know what the world will be like 26,000 years from now, but I'm pretty sure I won't be around to see it. Life is short.

OBSERVING TIPS

Face north. The easiest way to find Polaris is to use the Big Dipper. The two stars at the edge of the Dipper's "bowl" always point to the North Star. The North Star will be your latitude above the horizon, facing north. For example, if you live in Boston (42 degrees latitude) you'll see Polaris due north, 42 degrees above the horizon. At the North Pole (90 degrees latitude) you'll see Polaris straight overhead. At the equator (0 degrees latitude) it will be right at the horizon, and south of that, you'll never see it.

Star trails. With some practice and the right equipment you can point your camera at Polaris and get a beautiful shot of stars revolving around the pole. You'll need a steady tripod, a digital camera, and some software to combine multiple exposures. There are several tutorials on the web that explain how.

NORTH STAR REGION IN WINTER; 30-DEGREE FIELD OF VIEW

N

Polaris →

CEPHEUS

URSA MINOR

E

S

W

SWAN NEBULA (MESSIER 17)

BEAUTY: ★★★⤍
BRAGGING RIGHTS: A beautiful sight
HOW EASY IS IT TO SEE? Best with binoculars or small telescope
BEST TIME TO SEE IT: Summer (in Sagittarius)
TYPE: Diffuse Nebula
DISCOVERED: 1746 by Philippe Loys de Chéseaux

NOTES

For various physical and physiological reasons, our eyes see only a narrow slice of the electromagnetic spectrum: from low-frequency red light up to high-frequency blue/violet light. Of course, the universe doesn't care what frequency we can or cannot see; it is perfectly happy to spew out light at whatever frequencies it wishes. Black holes emit x-rays and gamma rays from their accretion disks, while newborn stars shrouded in dusty cocoons glow in infrared light.

Fortunately, humans are inventive creatures and we've developed tools to see these heretofore invisible frequencies. The best examples are NASA's four "Great Observatories." The Hubble Space Telescope is by far the most famous. It takes amazing pictures in visible light. Two others, the Compton Observatory and the Chandra Observatory, see gamma rays and x-rays respectively—light that is much higher in frequency than what we can see. The fourth is the Spitzer Space Telescope, and it is designed to see infrared radiation. When Spitzer pointed its camera at the Swan Nebula, it captured vast clouds of dust lit up by the glow of newborn stars—exactly how you'd expect a nebula to look. But when you compare Spitzer's image to its visible light counterpart, they are almost like photo-negatives: the bright parts of the infrared image are the dark parts of the visible light one.

The bright part of the Swan Nebula is surrounded by a dark shroud of dust, which hides the activity within—at least to the human eye. But Spitzer can pierce through the veil and reveal star factories churning out new suns. When you look at the Swan Nebula and marvel at its beauty, try to imagine how much of the world is right in front of you but hidden, completely invisible to the eye.

OBSERVING TIPS

M8's neighbor. The Swan Nebula (M17) is easy to find. Start at M8 (page 100) in Sagittarius with binoculars and scan up until you see M24 (page 132). Keep going a little farther up and you'll see what looks like a ghostly checkmark. That's the Swan Nebula.

Power up. Spend some time looking at M17 with low power; averted vision helps to reveal its true extent; it is larger than it initially seems. Then switch to medium or high power to see some of the detail in the bright areas.

SAGITTARIUS REGION IN SUMMER; 30-DEGREE FIELD OF VIEW

N

Eagle Nebula

Swan Nebula

M24

Trifid Nebula

M21

M22

Theta Ophiuchi

Antares

Lagoon Nebula (M8)

M19

E

SAGITTARIUS

SCORPIUS

S

M6

W

M7

PINWHEEL GALAXY (MESSIER 33)

BEAUTY: ★★★✦
BRAGGING RIGHTS: A beautiful sight
HOW EASY IS IT TO SEE? Best with binoculars or small telescope
BEST TIME TO SEE IT: Fall (in Triangulum)
TYPE: Galaxy
DISCOVERED: 1764 by Charles Messier

NOTES

When Charles Messier discovered Messier 33, he had no idea how far away it was. Today, we know it is almost 3 million light-years away. It's a discovery due, almost entirely, to the persistence and skills of a woman named Henrietta Leavitt.

Imagine I placed a 100-watt lightbulb some unknown distance away from you. Could you figure out how far away it is without moving? Sure! First you'd measure the light of a 100-watt lightbulb at a known distance—let's say 10 feet. Then, you would compare the light of the unknown bulb to your "standard." Using some basic algebra—light decreases with the square of the distance—you can calculate the distance to the unknown lightbulb. Until Henrietta Leavitt, however, we did not know of any "standard lights" in the universe. Leavitt was born in an era when women were not supposed to be scientists. Nevertheless, she persevered and discovered a type of star known as a Cepheid variable. These stars fluctuate in brightness in a well-known pattern. Leavitt discovered that their peak brightness depended on how long it took them to go from dim to bright.

This is like being able to determine the wattage of a bulb by timing how long it takes to reach full brightness. By measuring the distance to nearby Cepheids (via parallax), Leavitt could then calculate the distance to any visible Cepheid anywhere in the universe!

OBSERVING TIPS

Andromeda's little sister. Messier 33 is often overshadowed by its much larger and brighter neighbor: the Andromeda Galaxy (page 72). You can find M33 easily by finding the Andromeda Galaxy and then scanning down to Beta Andromedae. Keep going on the same line and you'll run into Messier 33.

Large but faint. If it were concentrated into a point, M33 would be a moderately bright star. Unfortunately, its light is spread out over an area equal to two full moons side by side. Low power, dark skies, and patience are required to see it.

NGC 604. M33 is host to one of the largest star-forming nurseries known. If it were as close to us as the Orion Nebula, it would fill our sky. But at millions of light-years distance, you can see this nebula as a fuzzy star near M33's nucleus.

ANDROMEDA REGION IN FALL: 30-DEGREE FIELD OF VIEW

N

M110

Andromeda Galaxy

M32

ANDROMEDA

Beta Andromedae

TRIANGULUM

M33

E

S

W.

PEGASUS

PLUTO

BEAUTY: ★↲
BRAGGING RIGHTS: A rare sight
HOW EASY IS IT TO SEE? Telescope required
TYPE: Dwarf planet
DISCOVERED: 1930 by Clyde Tombaugh

NOTES

In 1929, the solar system made sense. Four rocky planets hugged the sun in the inner system; four gaseous giants commanded the cold, outer system; and a belt of asteroids orbited in between. But then Pluto was discovered.

Pluto is an oddball. Its orbit is tilted with respect to the other planets—almost as if there wasn't enough room to fit it in. Instead of staying in its own lane, it sometimes crosses inside of Neptune's orbit. And it is a tiny rocky planet dancing in the realm of gas giants.

Perhaps Pluto's oddness is why we are so fond of it. It's an underdog, holding its own, and proud of being what it is without apology. No wonder there was such an outcry when the International Astronomical Union (IAU) decided Pluto would no longer be considered a planet.

But in a way, the IAU had no choice. In 1992, astronomers found another rocky object beyond Neptune. Though much smaller than Pluto, the discovery meant that Pluto had company. Since then, thousands of objects have been discovered sharing Pluto's territory. Indeed, Pluto is just the largest member of a group of oddballs known as the Kuiper Belt: worlds of rock and ice orbiting beyond Neptune. Astronomers believe hundreds of thousands of these objects remain undiscovered.

And thus Pluto is not so strange after all. Though it has lost its designation as a planet, it has gained an entire family.

OBSERVING TIPS

The edge of the solar system. Pluto is the most distant solar system object visible with amateur equipment. At its best, when Pluto is closest to Earth, it shines at magnitude 14—near the limit for an 8-inch telescope. Even then, you'll need dark skies and lots of patience. If you really want to spot it, head to a star party where you know a really big telescope will be present, as that'll help your chances.

Use a chart. You won't be able to distinguish Pluto from a star, even if you're looking right at it, so use a star chart and try to identify Pluto from its position in a star pattern. This won't be easy, as you'll constantly be flipping between the chart and the eyepiece.

Capture what you see. If you think you've found Pluto, try to draw the star pattern around it. Then, go back a week later and see if your Pluto has moved. If so, then you've got it right!

ZODIACAL LIGHT

BEAUTY: ★★
BRAGGING RIGHTS: An amazing sight
HOW EASY IS IT TO SEE? Very dark skies required
TYPE: Space dust
DISCOVERED: Seventeenth Century

NOTES

The solar system was born out of a spinning cloud of gas and dust. Gravity caused it to condense at the center, forming the sun, but the spinning cloud stayed in orbit and formed a flat disk, growing thinner with increasing distance.

Eventually, planets, moons, and countless asteroids formed from the disk, growing by gravitationally vacuuming up all the surrounding matter. Most worlds, including Earth, more or less stuck to the same orbital plane as the ancestral birth cloud. This is why all planets seem to travel along a narrow band in the sky—a band called the zodiac.

Though the original gas cloud is long gone, the flat plane of the solar system is not entirely empty. Dust from comets and debris from asteroid collisions are replenished out in space as fast as the planets can vacuum it. This collection of interplanetary dust forms a tenuous, flat cloud that also aligns with the zodiac.

Amazingly, this cloud is sometimes visible. At certain times of the year, when the zodiac is nearly perpendicular to the horizon, you can see a slight glow right after sunset or right before sunrise. This is zodiacal light—it is sunlight bounced off trillions of dust motes floating in the solar system.

And yet, if you were to fly in a spaceship to sample this dust it would be almost impossible. What appears from Earth to be a glowing cloud of dust is in reality so sparse that you've be hard pressed to detect anything but hard vacuum. Each dust mote in the zodiacal cloud is miles away from its nearest neighbor. You could fly for hundreds of miles through it without hitting a single particle.

OBSERVING TIPS

Equinox. Zodiacal light is best seen around the spring and autumn equinoxes. In spring, look at the western horizon after sunset; in autumn, look at the eastern horizon before sunrise. Needless to say, zodiacal light is extremely faint, so avoid the full moon and any kind of light pollution.

A triangle of light. Zodiacal light looks like a triangular patch of light with its base at the horizon. The tip points along the ecliptic plane—that is, the zodiac. How far up can you see the zodiacal light? The darker your skies, the more you'll be able to see.

OTHER SIGHTS

BEEHIVE CLUSTER (MESSIER 44)

BEAUTY: ★★★   **BRAGGING RIGHTS:** A beautiful sight
HOW EASY IS IT TO SEE? Best with binoculars or small telescope
BEST TIME TO SEE IT: Spring (in Cancer)
TYPE: Open Cluster   **DISCOVERED:** Known since antiquity

Finding the beehive. To find it, draw an imaginary line between Regulus (the brightest star in Leo) and Pollux (the left star in the Gemini Twins). Somewhere around halfway you'll see a little fuzzy mist.

A swarm of stars. Use binoculars to resolve the mist into a swarm of stars. The Beehive Cluster is more than two full moons across, so use low power to take it all in. This is one of the most beautiful open clusters in the sky, filled with bright stars, and large enough to view easily with binoculars.

SOMBRERO GALAXY (MESSIER 104)

BEAUTY: ★★★ **BRAGGING RIGHTS:** A beautiful sight
HOW EASY IS IT TO SEE? Best with a rich-field telescope
BEST TIME TO SEE IT: Spring (in Virgo)
TYPE: Galaxy **DISCOVERED:** 1781 by Pierre Méchain

A Virgo straggler? Astronomers believe this medium-size galaxy is part of the Virgo Cluster (page 166), though the association is not definitive. Messier 104 is around 30 million light-years away and is one of the brightest galaxies visible from Earth. To find it, start at Eta Corvi and move about 5½ degrees up and to the left toward Virgo.

Core and disk. No other galaxy shows the joining of a spherical core with a flat disk as clearly as Messier 104. Even in small telescopes you should be able to see a sharp line cutting across the central glow. In larger telescopes you might see a three-dimensional view: a soft glow below the dark edge causes the disk to pop forward. This is a beautiful galaxy in high-resolution photos, but even visually you should be able to see its general structure.

ESKIMO NEBULA (CALDWELL 39)

BEAUTY: ★★↙ **BRAGGING RIGHTS:** A beautiful sight
HOW EASY IS IT TO SEE? Best with a high-power telescope
BEST TIME TO SEE IT: Winter (in Gemini)
TYPE: Planetary Nebula   **DISCOVERED:** 1787 by William Herschel

Bright central star. Unlike the Ring Nebula (page 158), the central star of this planetary nebula is easily visible. In fact, you might have trouble seeing anything else at low power. To find it, start at Delta Geminorum and move to the left to 63 Geminorum. The Eskimo Nebula will be left and south from that.

Use high power. Like other bright but small objects, the Eskimo Nebula tolerates—and even requires—high power. Start at 75× or so to see the inner shell of the nebula. With averted vision, you might be able to see the outer shell also. The complex structure of the nebula is only visible in high-resolution photos, but if conditions allow, try boosting magnification up to ×150. You might be able to see a little more detail. Patience, practice, and lots of luck will reveal much.

MESSIER 87

BEAUTY: ★★↙ **BRAGGING RIGHTS:** A beautiful sight

HOW EASY IS IT TO SEE? Best with a small telescope

BEST TIME TO SEE IT: Spring (in Virgo)

TYPE: Galaxy **DISCOVERED:** 1781 by Charles Messier

Monster in Virgo. With a mass of 800 billion suns (twice that of the Milky Way) elliptical galaxy Messier 87 dominates the Virgo Cluster physically and visually. Though lacking detail even in high-resolution images, this galaxy impresses with its size and brightness.

To find it, star hop from Epsilon Virginis to Rho Virginis, and then to 20 Virginis. From there move about 2 degrees north to find M87 (see chart).

Black hole heart. But the real monster is invisible. A giant black hole—as heavy as a small galaxy—lurks in M87's heart. Gas and stars are ripped into plasma by the black hole's ferocious gravity. Magnetic field lines blast the ionized remains along a 5,000 light-year-long jet. It's like a dragon belching a stream of fire.

"I've seen it!"

TRIFID NEBULA (MESSIER 20)

BEAUTY: ★★↙   **BRAGGING RIGHTS:** A beautiful sight
HOW EASY IS IT TO SEE? Best with a rich-field telescope
BEST TIME TO SEE IT: Summer (in Sagittarius)
TYPE: Diffuse Nebula   **DISCOVERED:** 1764 by Charles Messier

Summer jewel. The summer skies around Sagittarius are filled with wonders. The Lagoon Nebula is, of course, the best of them (see page 100), but this little four-leafed clover of a nebula is worth seeking out. Move 1 degree north of the Lagoon Nebula to find M20.

Three or four lobes? Though beautiful in long-exposure photographs, the Trifid Nebula is faint and delicate when seen through an eyepiece; even in low power you'll have to use averted vision. Nevertheless, you should be able to see the dark lanes of gas that split it into four lobes. Three of the lobes are relatively easy to spot, while the fourth is always a challenge. Can you see it? When you're looking at M21, don't miss Messier 21, less than one degree north of the Trifid. Though they are not physically associated, they look beautiful together.

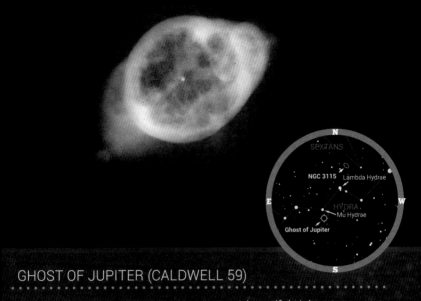

GHOST OF JUPITER (CALDWELL 59)

BEAUTY: ★★★ **BRAGGING RIGHTS:** A beautiful sight
HOW EASY IS IT TO SEE? Best with a high-power telescope
BEST TIME TO SEE IT: Spring (in Hydra)
TYPE: Planetary Nebula **DISCOVERED:** 1785 by William Herschel

What's in a name? Caldwell 59 is close in size to Jupiter (as seen from Earth) and shares its pale yellow color, hence the nickname, "Ghost of Jupiter." Perhaps the Millennial Generation will be reminded of a bright lidless eye and end up calling it "The Eye of Sauron Nebula."

Magnify! To find the Ghost of Jupiter, start at Mu Hydrae and move about 2 degrees south. You might see the Ghost of Jupiter with binoculars as a faint star, but it will be too small to reveal details. Fortunately, the nebula is bright enough to stand up to magnification. Magnification of 100× or more reveals its concentric shell. Look at the Ghost of Jupiter with averted vision to see the full extent of the nebula. When you look away, you'll see a lone star, the nebula's progenitor star—the one that blew its outer layers into space.

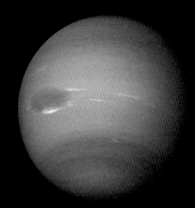

NEPTUNE

BEAUTY: ★★   **BRAGGING RIGHTS:** A rare sight
HOW EASY IS IT TO SEE? Telescope required
TYPE: Planet   **DISCOVERED:** Predicted 1846 by Urbain Le Verrier

At opposition. The best time to look for Neptune is when it is opposite the sun with respect to Earth (this is known as "in opposition"). When Neptune is in opposition, it'll rise at sunset and be high in the sky at midnight. From now until the late 2020s, Neptune will be in opposition during September; views in August and October will also be good.

Use a chart. Neptune is seldom brighter than the surrounding stars, and at low magnification it will be indistinguishable from a star, though you might notice its faint blue-green color. A bigger telescope captures more light and allows you to see more detail. At 200× you might be able to see Neptune's disk; more magnification will make it clearer. Unlike faint galaxies, which you need dark skies to observe, you should be able to see Neptune in a telescope even from a suburban backyard. But you need calm and steady skies to see it clearly.

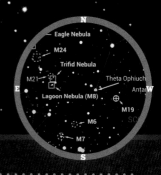

MESSIER 6 AND MESSIER 7

BEAUTY: ★★★ **BRAGGING RIGHTS:** A beautiful sight
HOW EASY IS IT TO SEE? Best with binoculars or small telescope
BEST TIME TO SEE IT: Summer (in Scorpius)
TYPE: Open Cluster **DISCOVERED:** Second Century by Ptolemy

The scorpion's tail. These two beautiful open clusters are near Lambda Scorpii—the stinger of Scorpius. The entire region around the scorpion's tail is filled with open clusters and faint nebulae.

Contrast. Start by finding Scorpius (page 35), then find both clusters with binoculars. Find Lambda Scorpii and pan a little to the left and up. M6 will be a tight group of stars off by itself, set against a relative dark part of the sky. By contrast, M7 will be a looser cluster competing against a sea of stars. Both are beautiful on their own, but the contrast is interesting too. A small telescope enhances M6. You'll see a smattering of blue stars—many people see the shape of a butterfly. At one edge, a bright red star appears like a ruby in a pile of diamonds. M7 is a larger cluster and best viewed at low magnification.

COMA STAR CLUSTER

BEAUTY: ★★★ **BRAGGING RIGHTS:** A beautiful sight
HOW EASY IS IT TO SEE? Just look up or use binoculars
BEST TIME TO SEE IT: Spring (in Coma Berenices)
TYPE: Open Cluster **DISCOVERED:** Known since antiquity

Naked-eye cluster. Start by appreciating this cluster with the naked eye. From a dark site, you can see a handful of stars at magnitude 5 in the Coma Berenices constellation (page 26). This is one of the closest clusters to Earth, so the stars appear spread out. The ancient Greeks identified this constellation as the end of the tail of Leo, the lion. Later, it honored the cut-off hair of Egyptian Queen Berenice—offered to the gods by her for the safe return of her husband from war.

Binocular view. Train your binoculars on this cluster and 50 more stars will appear. The Coma Star Cluster, also known as Melotte 111, is around 280 light-years away from Earth—twice as far as the Hyades cluster (page 130).

MESSIER 11

· ·

BEAUTY: ★★★ **BRAGGING RIGHTS:** A beautiful sight
HOW EASY IS IT TO SEE? Best with a small telescope
BEST TIME TO SEE IT: Summer (in Scutum)
TYPE: Open Cluster **DISCOVERED:** 1681 by Gottfried Kirch

Small but rich. Messier 11 is only about 13 arcminutes across—less than half the width of the full moon—yet it packs hundreds of stars. Under dark skies you may be able to see this cluster with the naked eye as a faint fuzzy blob. Look halfway between Alpha Scuti and Lambda Aquilae. With binoculars the blob resolves into scores of faint stars.

Use moderate power. A telescope resolves more stars. Can you see a V shape in the stars? This cluster is sometimes called the "Wild Duck Cluster" because of this birdlike shape. Use moderate power (50× or so) to resolve the stars at the center of the cluster. Messier 11 is in a rich and bright part of the Milky Way. Sweep around and get lost in this sea of stars. Under very dark skies, you might be able to see various nebulae.

"I've seen it!"

HELIX NEBULA (CALDWELL 63)

BEAUTY: ★★★   **BRAGGING RIGHTS:** A beautiful sight
HOW EASY IS IT TO SEE? Best with a rich-field telescope
BEST TIME TO SEE IT: Fall (in Aquarius)
TYPE: Planetary Nebula
DISCOVERED: 1862 by Georg Friedrich Julius Arthur von Auwers

Large but faint. The ring-shaped Helix Nebula is brighter than the Ring Nebula (page 158) but all those photons are spread out over a much larger area. The result is that it looks much fainter, at least to our eyes. To find it, start at Fomalhaut and star hop to Upsilon Aquarii, then look a little more than 1 degree to the right (see chart). Use binoculars or a low-power (rich-field) telescope to concentrate the light. Even then, you may have trouble finding this object under light-polluted suburban skies. If so, try again under the dark skies of a National Park. Like all faint fuzzies, patience and persistence are required. After you've found it, spend time looking at it. Alternate between direct and averted vision. The inner ring is the brightest part. You may only see a partial circle at first, but keep looking and see if you can make it whole.

○ *"I've seen it!"*

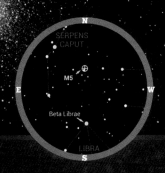

MESSIER 5

BEAUTY: ★★★   **BRAGGING RIGHTS:** A beautiful sight
HOW EASY IS IT TO SEE? Best with a small telescope
BEST TIME TO SEE IT: Summer (in Serpens)
TYPE: Globular Cluster   **DISCOVERED:** 1702 by Gottfried Kirch

M13's rival? Messier 5 is one of the best globular clusters in the Northern Hemisphere. Messier 22 (page 134) is bigger and brighter, and the more famous Hercules Cluster (page 112) is not necessarily superior.

A blur in binoculars. Though easily visible in binoculars, you won't see much more than a faint glow. To find it, start at Beta Librae and move 10 degrees north (about a fist held at arm's length). Charles Messier, using a small telescope, thought it was a nebula. But a good telescope at 100× reveals a swarm of stars, though much of the center will remain unresolved. Famed observer Stephen O'Meara reports seeing color at low power that he described as "a straw interior with a powder blue exterior." What do you see?

"I've seen it!"

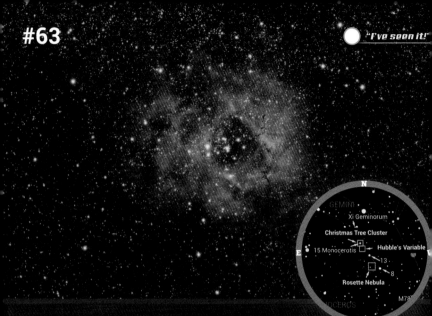

ROSETTE NEBULA (CALDWELL 49)

BEAUTY: ★★★ **BRAGGING RIGHTS:** A beautiful sight
HOW EASY IS IT TO SEE? Best with a rich-field telescope
BEST TIME TO SEE IT: Winter (in Monoceros)
TYPE: Diffuse Nebula and Open Cluster
DISCOVERED: 1784 by William Herschel

Cluster and nebula. The Orion Nebula is young: its stars are still enclosed in their birth cocoon. The Pleiades are older: they've already drifted far away from their birthplace. But the Rosette Nebula is in between—its bright stars shine gloriously against the glow of their "nursery" nebula. To find it, start at Hubble's Variable Nebula (page 213) and then follow the line from 13 to 8 Monocerotis. You'll see the cluster about 2 degrees left of 8 Monocerotis. Once you've found the cluster, patiently try to detect the nebula with averted vision. When looking at the nebula, you will see patches and tendrils; these are the brightest parts of the nebula. Try using a nebula filter for a better look; such filters block light pollution but preserve views of the nebula.

◯ *"I've seen it!"*

SUPERMOON

BEAUTY: ★★★ **BRAGGING RIGHTS:** A beautiful sight
HOW EASY IS IT TO SEE? Just look up
TYPE: Special event **DISCOVERED:** First hyped 1979

Super-pseudoscience? With origins in astrology, discredited theories about earthquakes, and media hype, the term "supermoon" is sometimes avoided by astronomers. But any excuse to look at the sky is good, in my opinion, as long as you know what you're seeing. A supermoon happens when the moon is closest to Earth in its orbit (at perigee) and is simultaneously is in its full phase. When this happens, the full moon appears slightly bigger and brighter. Even so, judging the size of the moon is difficult. When the moon is close to the horizon it looks bigger to us than when it is high in the sky—probably because we're comparing it against mountains and buildings. (This "Moon Illusion" has a bigger impact than the 14 percent increase in size due to a supermoon.) On supermoon day, look to the east at sunset to watch the moon rise. Seeing it against the horizon, maybe reddened by the intervening atmosphere, is a great way to get ready for the show.

CAT'S EYE NEBULA (CALDWELL 6)

BEAUTY: ★★★   **BRAGGING RIGHTS:** A beautiful sight
HOW EASY IS IT TO SEE? Best with a high-power telescope
BEST TIME TO SEE IT: Summer (in Draco)
TYPE: Planetary Nebula   **DISCOVERED:** 1786 by William Herschel

As seen by Hubble. The Hubble Space Telescope's image of the Cat's Eye Nebula is justly famous. It shows complex and multilayered shells that look much different than the more symmetrical shells of the Helix Nebula or the Ring Nebula. While very little of that detail is visible to the eye—even in large telescopes—keep the Hubble photo in mind when looking at this beautiful object. To find the Cat's Eye Nebula, look about halfway between Delta and Zeta Draconis.

Use high magnification. This tiny nebula—about the size of Mars as seen from Earth—needs lots of magnification for a proper view. 200× or more is appropriate, as long as steady skies allow it. And of course, larger telescopes have better resolution and can enlarge the view more effectively.

SCULPTOR GALAXY (NGC 253)

BEAUTY: ★★★ **BRAGGING RIGHTS:** A beautiful sight
HOW EASY IS IT TO SEE? Best with a rich-field telescope
BEST TIME TO SEE IT: Fall (in Sculptor)
TYPE: Galaxy **DISCOVERED:** 1783 by Caroline Herschel

Herschel's galaxy. Caroline Herschel was William Herschel's sister, and this is one of at least a dozen deep-sky objects she discovered. But the Herschels were observing from England, and this southern galaxy is never far from the hazy horizon. It was William Herschel's son John, observing from South Africa, who had a chance to see its "streaky and knotty" appearance, and who speculated that it might be resolvable into stars. From North America you'll need a clear view of the southern horizon to see this amazing galaxy. If it were higher up in the sky it would be much more famous, as it is brighter than all but a few others. To find it, look 7½ degrees south of Beta Ceti. Many observers report seeing a mottled look, particularly around the galaxy's core. With dark skies you might also see two faint galactic arms sticking out to either side.

○ "I've seen it!"

CHRISTMAS TREE CLUSTER (NGC 2264)

BEAUTY: ★★★ **BRAGGING RIGHTS:** A beautiful sight
HOW EASY IS IT TO SEE? Best with a rich-field telescope
BEST TIME TO SEE IT: Winter (in Monoceros)
TYPE: Diffuse Nebula and Open Cluster
DISCOVERED: 1784 by William Herschel

Hidden treasure. This beautiful cluster is neither in the Messier nor Caldwell catalogs, leading Stephen O'Meara to dub it a "hidden treasure." As in all overlooked objects, there is joy in discovering it; when it comes to the Christmas Tree Cluster, the sparkle of more than 30 bright stars (conveniently shaped like a Christmas tree) is quite a treat. Of course, there is one main reason it was overlooked in the first place: the surrounding nebula is quite dim—far fainter than the Rosette Nebula (page 196)—diminishing its rank on our list. Still, it should still be visible with the naked eye in very dark skies. Binoculars show the stars clearly. The cluster is relatively large, bigger than the full moon, so use low magnification to see it all. To find it, look for the fifth-magnitude star 15 Monocerotis (see chart), the brightest star in NGC 2264.

"I've seen it!"

NORTH AMERICA NEBULA (CALDWELL 20)

BEAUTY: ★★★ **BRAGGING RIGHTS:** A beautiful sight

HOW EASY IS IT TO SEE? Best with a rich-field telescope

BEST TIME TO SEE IT: Fall (in Cygnus)

TYPE: Diffuse Nebula   **DISCOVERED:** 1786 by William Herschel

A continent of stars. With apologies to our Alaskan and Hawaiian friends, this nebula bears a striking resemblance to the contiguous part of the North American continent. It is also surprisingly easy to find. Point your binoculars at Deneb, the brightest star in Cygnus, and pan to the left (east) a few degrees. Under dark skies, you should see a fan of nebulosity more than a degree across (twice the diameter of the full moon). You could spend hours exploring the patches of nebula and small star clusters that make up this nebula. An arc of light resembling Mexico is one of its most prominent pieces. From there, you might be able to make out the rest of the Gulf Coast and a faint peninsula representing Florida. Two small open clusters (NGC 6997 and Collinder 428) are embedded in the nebula, the first on the eastern side and the second on the west.

MESSIER 83

BEAUTY: ★★★ **BRAGGING RIGHTS:** A beautiful sight
HOW EASY IS IT TO SEE? Best with a rich-field telescope
BEST TIME TO SEE IT: Spring (in Hydra)
TYPE: Galaxy   **DISCOVERED:** 1752 by Nicolas-Louis de LaCaille

Low in the southern skies. Viewers north of Philadelphia (~40 degrees latitude) might have trouble seeing this beautiful galaxy. As seen from the City of Brotherly Love, Messier 83 never rises more than 20 degrees above the southern horizon—less than the distance between your thumb and pinkie with your arm stretched out. Nevertheless, Messier himself saw the galaxy (barely) from Paris, which is north of almost all of the continental United States, so try to see it anyway. To find this galaxy, first locate a small group of stars about 20 degrees south of Spica (see chart). These stars, 1, 2, and 3 Centauri, are only 3 degrees south of the galaxy. Pan up and try to find it. Were it not for its southern location, Messier 83 would be one of the highlights of the sky. Only a handful of galaxies are brighter, so don't pass up the chance to see it if the southern horizon is unobstructed and the area's free of most light pollution.

MESSIER 4

BEAUTY: ★★★ **BRAGGING RIGHTS:** A beautiful sight
HOW EASY IS IT TO SEE? Best with binoculars or small telescope
BEST TIME TO SEE IT: Summer (in Scorpius)
TYPE: Globular Cluster
DISCOVERED: 1746 by Philippe Loys de Chéseaux

In the glare of Antares. Find this beautiful globular cluster on summer days when Scorpius is high in the south (page 35). With binoculars or a small telescope, point to Scorpius's brightest star, Antares, then slew a little more than a degree west to find Messier 4. Though its proximity to Antares makes it easy to find, the red giant's glare washes out the cluster's light. Unlike the tight cores of Messier 5 and Messier 13, this cluster has a looser core. Using higher power and larger scopes brings out more stars, but even small instruments can reveal much detail. Two other globular clusters are nearby. Messier 80, a fainter but tighter globular, is about three degrees away to the northwest. In the other direction, much closer to Antares, you'll find NGC 6144; moderate telescopes (and experienced observers) should be able to spot it. Can you see it?

"I've seen it!"

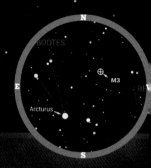

MESSIER 3

BEAUTY: ★★★ **BRAGGING RIGHTS:** A beautiful sight
HOW EASY IS IT TO SEE? Best with a small telescope
BEST TIME TO SEE IT: Spring (in Canes Venatici)
TYPE: Globular Cluster **DISCOVERED:** 1764 by Charles Messier

Cluster near Arcturus. Messier 3 is relatively easy to find: In late spring or early summer, look up in the sky and find Arcturus, a bright yellow-orange star; to do so, trace the arc of the Big Dipper's handle to Arcturus. Use binoculars to scan the area. Six degrees away from Arcturus, in a relatively empty area of the sky, you'll find a fuzzy glow surrounded by tiny stars! Some people consider Messier 3 to be the second-best globular cluster in the sky—after M13 (page 112). Personally, I think M22 (page 134) and M5 (page 195) are better. M5 and M13 are visible around the same time of the year, so you should be able to compare them to M3 and make up your own mind. Unlike galaxies and nebulae, globular clusters can stand high power. Try increasing the magnification to see if you can resolve more of the

stars near the nucleus.

"I've seen it!"

North America Nebula
Deneb
CYGNUS
Epsilon Cygni
Veil Nebula ε2

VEIL NEBULA (CALDWELL 33 AND 34)

BEAUTY: ★★✦   **BRAGGING RIGHTS:** A beautiful sight
HOW EASY IS IT TO SEE? Best with a rich-field telescope
BEST TIME TO SEE IT: Fall (in Cygnus)
TYPE: Supernova Remnant
DISCOVERED: 1784 by William Herschel

Tomb of an unknown star. Unlike the famous Crab Nebula (page 126), there are no records marking the death of the star that spawned the Veil Nebula. Its star exploded at least 5,000 years ago, before the invention of writing. The gaseous remains of the supernova have expanded since then, now encompassing an area several times wider than the full moon (as seen from Earth). Different fragments in the nebula shine brighter than others, and they have their own names. Caldwell 33 (also known as NGC 6992) is the bright Eastern Veil, while Caldwell 34 (NGC 6960) is the fainter Western Veil. To find the nebula, start at Epsilon Cygni and then move about 3 degrees south to 52 Cygni. If your skies are dark enough, you should see the Western Veil touching 52 Cygni. Using binoculars or a rich-field telescope will help, as will dark skies and using averted vision. **205**

MESSIER 101

BEAUTY: ★★☆ **BRAGGING RIGHTS:** A beautiful sight

HOW EASY IS IT TO SEE? Best with rich-field telescope

BEST TIME TO SEE IT: Summer (in Ursa Major)

TYPE: Galaxy **DISCOVERED:** 1781 by Pierre Méchain

Faint giant. Messier 101, the famous Pinwheel Galaxy, looks beautiful in astrophotos, like a swirling maelstrom of starlight. Unfortunately, the view through an eyepiece is less dramatic. Though bright, Messier 101's light is spread out over a large area and even the mildest light pollution obliterates its faint arms. To find it, start at the visual double in the handle of the Big Dipper, Alcor and Mizar. Follow the line of stars forking away from the handle and you should see the faint glow of M101. If conditions allow, however, you're in for a treat. Start at low power and find the fuzzy core. Do you see a faint oval expanse around the starlike core? With larger telescopes and dark skies you might be able to trace out one or more of the arms. Most of the visible detail is in the core. Use moderate power (50× to 100×) to tease out the central core. This is one of the more challenging objects on the list, but it's worth it.

MESSIER 106

BEAUTY: ★★✶ **BRAGGING RIGHTS:** A beautiful sight
HOW EASY IS IT TO SEE? Best with rich-field telescope
BEST TIME TO SEE IT: Spring (in Canes Venatici)
TYPE: Galaxy **DISCOVERED:** 1781 by Pierre Méchain

Cuttlefish in space? This is one of the first galaxies I photographed, and when I saw the result I thought it looked like the body of a squid or a cuttlefish. Its bright central area is oval and mottled, and much more prominent than its thin arms. To find it, start at Gamma Ursa Majoris (Phecda) and move about 6 degrees in the direction of Alpha Canum Venaticorum.

A violent nucleus. A monster black hole lurks at the heart of Messier 106. It devours stars whole and spews out beams of high-energy particles from its poles. Most of the violence is invisible to us, but the brightness of the nucleus is a clue that a lot of energy is in play.

Focus on the core. Most of the detail in this galaxy is found near the core. Use moderate power to unravel its patchy surroundings.

MESSIER 35

BEAUTY: ★★★ BRAGGING RIGHTS: A beautiful sight
HOW EASY IS IT TO SEE? Best with small telescope
BEST TIME TO SEE IT: Winter (in Gemini)
TYPE: Open Cluster **DISCOVERED:** 1745, Philippe Loys de Chéseaux

At the foot of the twin. Messier 35 is easy to find. Find Gemini in late winter and look at the foot of the right-most twin (from our perspective). A few degrees away you'll see a faint concentration of stars, just a fuzzy patch of light. Though it is (barely) visible to the naked eye, optical aid is required to best appreciate this cluster of faint stars. Binoculars show a few dozen outlying stars buzzing around a faint unresolved core. A small telescope at 50× shows more stars, though patience and averted vision are required to see them all. For an added challenge, try spotting NGC 2158, which is a tiny open cluster about a half degree to the south. In a small telescope it will appear as a small fuzzy orb. Larger scopes can resolve it into faint stars.

MESSIER 15

BEAUTY: ★★★ BRAGGING RIGHTS: A beautiful sight
HOW EASY IS IT TO SEE? Best with small telescope
BEST TIME TO SEE IT: Fall (in Pegasus)
TYPE: Globular Cluster
DISCOVERED: 1746, Jean-Dominique Maraldi II

Collapsed core. To blast a satellite into orbit you need to give it lots of kinetic energy, usually by putting on top of a rocket. The greater the speed, the higher the satellite can go. The stars in a globular cluster are all orbiting with enough energy to keep them buzzing around like bees. But sometimes, those orbits can decay. In that case, the stars fall toward the center of the cluster, clumping together. This is known as a collapsed core, and Messier 15 is a perfect example of this. In fact, M15 is one of the densest clusters known—thousands of stars are packed into a few cubic light-years. Look for M15 about 4 degrees west-northwest of Epsilon Pegasi. M15's bright but small core is visible with binoculars, but you'll need a telescope at moderate power to resolve its outlying stars. Even then, you'll need a large telescope to spot the core. **209**

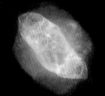

GHOST OF SATURN (CALDWELL 55)

BEAUTY: ★★★ **BRAGGING RIGHTS:** A beautiful sight
HOW EASY IS IT TO SEE? Best with a high-power telescope
BEST TIME TO SEE IT: Fall (in Aquarius)
TYPE: Planetary Nebula **DISCOVERED:** 1782 by William Herschel

Complex structure. This beautiful planetary nebula is known as the "Ghost of Saturn" because it has a glowing bar across its equator, which resembles Saturn's rings when seen edge-on. Unfortunately, such detail is invisible in most backyard telescopes (though a skilled astrophotographer can capture it). To find this ghost, look about 1½ degrees west (to the right) of Nu Aquarii (see chart).

Another eye? Instead of a Saturn-like scene, you'll see is something resembling the Cat's Eye Nebula (page 198): a bright central star surrounded by a faint oval ring of nebulosity. If steady skies allow it, use high magnification (×100 or more) to see detail in this tiny nebula. The nebula appears no larger than the planet Saturn, so you'll need significant magnification.

MESSIER 37

BEAUTY: ★★✦ BRAGGING RIGHTS: A beautiful sight

HOW EASY IS IT TO SEE? Best with a small telescope

BEST TIME TO SEE IT: Winter (in Auriga)

TYPE: Open Cluster

DISCOVERED: Before 1654 by Giovanni Batista Hodierna

Use a telescope. Messier 37 is easily visible in binoculars—from a dark site it should be visible to the naked eye—but the cluster is so small that it can't be appreciated at low magnification. To find it, start at Theta Aurigae and pan south (down) about 5 degrees (see chart). Use moderate power (50× to 100×) to dive into this huge collection of stars. Like a sparse globular cluster: Messier 37 has almost 2,000 members, enough to make it seem like a sparse globular cluster.

Though not as beautiful, two other open clusters are only 5 degrees away to the northwest. M36 has fewer stars but is slightly more compact. M38 is like a fainter version of M36, though it is almost as large as M37.

RS OPHIUCHI

BEAUTY: ★★ **BRAGGING RIGHTS:** A rare sight
HOW EASY IS IT TO SEE? Telescope required
BEST TIME TO SEE IT: Summer (in Ophiuchus)
TYPE: Recurring Nova **DISCOVERED:** 1904 by Williamina Fleming

The odd couple. RS Ophiuchi isn't much to look at, a faint star in a field of hundreds, barely visible in a telescope. It's actually a binary system in which a red giant star is locked in a gravitational embrace with a tiny white dwarf. Like a vampire sucking blood, the incredibly dense white dwarf siphons hydrogen gas from the red giant. Bit by bit the gas accumulates around the white dwarf until it reaches a critical mass. Then, atomic nuclei suddenly fuse, releasing huge amounts of energy. The resulting blast is visible thousands of light-years away, all the way to Earth, where the star becomes visible to the unaided eye. Though not as bright as a supernova, its eruptions are more frequent. On average they happen every 20 years. The last eruption was in 2006; during its quiet period, it's difficult to find, but every 20 years, you'll see a bright star where one didn't exist before.

HUBBLE'S VARIABLE NEBULA (CALDWELL 46)

BEAUTY: ★★ **BRAGGING RIGHTS:** A beautiful sight
HOW EASY IS IT TO SEE? Telescope required
BEST TIME TO SEE IT: Winter (in Monoceros)
TYPE: Diffuse Nebula **DISCOVERED:** 1783 by William Herschel

Hidden complexity. Most nebulae change slowly, if at all, but this object, a small fan-shaped nebula in the Monoceros constellation, changes from month to month, like a slow-motion film of a flaming torch. We don't know for sure what's causing the phenomena, but astronomers believe a binary star is responsible. A large star—10 times heavier than our sun—is pulling gas from its companion, and emitting fierce jets of gas and radiation as it consumes its meal. Filaments of gas spew out—possibly aligned on magnetic field lines—and cast moving shadows visible all around the nebula. To find it, start at Xi Geminorum and move down to 15 Monocerotis, which is part of the Christmas Tree Cluster (page 200). Continue moving on the same line another 1¼ degrees and you'll reach Hubble's Nebula. The nebula is faint, however, so dark skies and averted vision are required.

"I've seen it!"

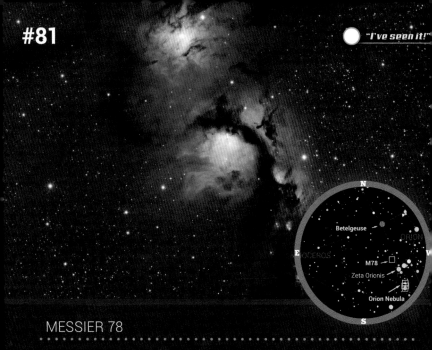

MESSIER 78

BEAUTY: ★★ **BRAGGING RIGHTS:** A beautiful sight
HOW EASY IS IT TO SEE? Best with a rich-field telescope
BEST TIME TO SEE IT: Winter (in Orion)
TYPE: Diffuse Nebula **DISCOVERED:** 1780 by Pierre Méchain

The Lesser Orion Nebula. Though its neighbor, Messier 42 (page 92), rightly overshadows this nebula, Messier 78 has many charms worth exploring. To find it, start at Zeta Orionis—the left-most star in Orion's belt, and move northeast about 2½ degrees (see chart). At low power it is a fuzzy, extended star, almost like a comet. With more magnification you can see two or three concentrations of light toward its center. With averted vision and dark skies you might even be able to see dark lanes crossing the nebula.

The nebulous neighborhood. Messier 78 is the brightest nebula in a patchwork of faint objects. At least three other small patches of light are visible with a telescope. Use a rich-field scope to explore the neighborhood and see how many other patches of light you can detect.

MESSIER 77

BEAUTY: ★★✦ **BRAGGING RIGHTS:** A beautiful sight
HOW EASY IS IT TO SEE? Best with rich-field telescope
BEST TIME TO SEE IT: Winter (in Cetus)
TYPE: Galaxy **DISCOVERED:** 1780 by Pierre Méchain

M106's twin? Compare astrophotos of Messier 77 with Messier 106. If you tilted M77 at an angle, wouldn't it look like M106? Both share a bright, chaotic inner region and a very faint and tenuous set of outer arms. In binoculars, M77 looks like a faint, possibly fuzzy star. It is 1 degree away from Delta Ceti (see chart). A telescope reveals two zones: an inner, starlike core, and a fainter disk region. Long-exposure photographs show an even fainter set of outer arms, but these are not visible through amateur telescopes.

Use moderate magnification (75× or so) to examine the fainter disk. Do you see any detail? Early observers, not knowing how far away this object is, saw concentrated knots and thought they were stars. We know better, however, and realize that they are entire star-forming regions.

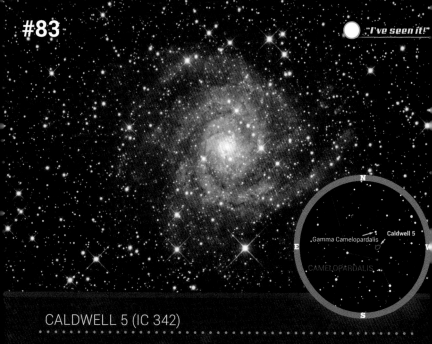

N

Caldwell 5

Gamma Camelopardalis

E

W

CAMELOPARDALIS

S

CALDWELL 5 (IC 342)

BEAUTY: ★★✦ **BRAGGING RIGHTS:** A beautiful sight
HOW EASY IS IT TO SEE? Best with a rich-field telescope
BEST TIME TO SEE IT: Winter (in Camelopardalis)
TYPE: Galaxy **DISCOVERED:** ~1890 by William F. Denning

Face-on spiral. Face-on spirals like Messier 51 (page 168) and Messier 101 (page 206) are challenging objects: their light is spread out over a large area, making them vulnerable to even a hint of light pollution or haze. Caldwell 5 is no better, but it is also no worse, which raises the question of why it isn't more famous. Famed observer Stephen O'Meara even says that this galaxy's spiral structure is "much easier to trace than either M74 or M101." Find this object—just 3¼ degrees south of Gamma Camelopardalis—and decide whether or not he's right.

Use low power. The galaxy is big and you'll need low power to concentrate its light. Start at 25× and look at the core. Can you see any fuzziness with averted vision? If so, try to trace it out away from the core.

○ *"I've seen it!"*

MESSIER 41

BEAUTY: ★★☽ **BRAGGING RIGHTS:** A beautiful sight
HOW EASY IS IT TO SEE? Best with rich-field telescope
BEST TIME TO SEE IT: Winter (in Canis Major)
TYPE: Open Cluster   **DISCOVERED:** 1702 by John Flamsteed

The Little Beehive? Messier dismissed this object as "no more than a cluster of faint stars," but other observers call it the "Little Beehive," for its resemblance to the spectacular Messier 44 (page 184). Perhaps the conflict can be resolved by noting that the cluster is larger than the full moon and Messier observed at high powers, which would have made it harder to see.

Naked-eye view. At magnitude 4.5, this cluster should be easily visible with the naked eye—look 4 degrees south of Sirius. Binoculars, however, should show the cluster clearly and resolve it into several stars. After you spot it, increase the magnification to 25× power. Can you see any color in the stars? Many people see a few yellow-orange stars in a field of icy blue ones. The central star is the brightest and most colorful.

MESSIER 2

BEAUTY: ★★⭐   **BRAGGING RIGHTS:** A beautiful sight
HOW EASY IS IT TO SEE? Best with a high-power telescope
BEST TIME TO SEE IT: Fall (in Aquarius)
TYPE: Globular Cluster
DISCOVERED: 1746 by Jean-Dominique Maraldi II

Lonely cluster. Look at the constellation Aquarius on a crisp fall night and you'll be looking away from the crowded Milky Way, and out toward the vast nothingness of intergalactic space. Messier 2 shines proudly in this darkness, with only a few faint stars for company (see chart). Its isolation enhances the view: the cluster is an oasis of light in a dark desert. At low power, Messier 2 looks like a fuzzy star, but magnification resolves more stars and brings out the details. At 100× you'll see dozens of faint stars against a concentrated haze of even fainter, unresolved stars. Use averted vision for best results. If skies and equipment allow, try increasing the magnification even more.

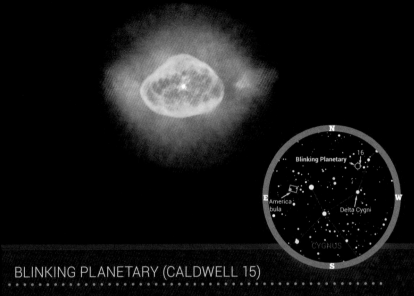

BLINKING PLANETARY (CALDWELL 15)

BEAUTY: ★★↗ **BRAGGING RIGHTS:** A beautiful sight
HOW EASY IS IT TO SEE? Best with a high-power telescope
BEST TIME TO SEE IT: Summer (in Cygnus)
TYPE: Planetary Nebula **DISCOVERED:** 1793 by William Herschel

Averted vision. The Blinking Planetary consists of two concentric shells: a bright inner shell and a larger, but dimmer, outer shell. When you look at it with direct vision, the bright inner shell is visible, but the outer shell is too faint to see. When you look away, the nebula seems to grow in size. Try it!

Find this enigma next to 16 Cygni, which is itself about 6 degrees north of Delta Cygni (see chart). The Blinking Planetary is tiny, and at less than 50× you may not see much more than a star. Try 100× to see more of the outer shell. If your skies and telescope allow, keep increasing the magnification and see if you can detect any detail.

"I've seen it!"

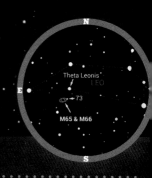

MESSIER 65 AND MESSIER 66

BEAUTY: ★★↗ BRAGGING RIGHTS: A beautiful sight
HOW EASY IS IT TO SEE? Best with a rich-field telescope
BEST TIME TO SEE IT: Spring (in Leo)
TYPE: Galaxy **DISCOVERED:** 1780 by Pierre Méchain

Double or triple. M65 and M66 form a famous pair of galaxies—only 20 arcseconds apart—and they are second only to M81 and M82 in popularity and beauty (see page 104). To find them, start at Theta Leonis and then move south to 73 Leonis. The pair (or triplet) should be about a full moon's width to the left (east-southeast).

Messier 65 & 66. M65 is technically slightly fainter overall than M66, but since it is more compact, it may appear brighter. This galaxy appears nearly edge on; you should be able to see it as an oval glow. Not much detail is visible within its disk, though you should notice its bright center. In photos, M66 shows two strong, prominent arms; the visual view is much less impressive, unfortunately. The bright nucleus is easily seen, but the arms are revealed only through large telescopes.

PAC-MAN NEBULA (NGC 281)

BEAUTY: ★★✦ **BRAGGING RIGHTS:** A beautiful sight
HOW EASY IS IT TO SEE? Best with a rich-field telescope
BEST TIME TO SEE IT: Fall (in Cassiopeia)
TYPE: Diffuse Nebula **DISCOVERED:** 1881 by E. E. Barnard

Barnard's Nebula. Like Messier a century before him, E. E. Barnard hunted comets (for money) and cataloged comet-impostors like this beautiful nebula. He discovered Barnard's Star, the second-closest star to the sun, and he created a catalog of dark nebulae, now named Barnard 1 through Barnard 370. Under dark skies, this nebula is visible in binoculars as a faint glow in Cassiopeia, about 1½ degrees east of Alpha Cassiopeia. In a telescope, use low power to see it—the nebula is about the size of the full moon. Even then, you'll need averted vision and minimal light pollution. A light-pollution filter or a nebula filter might help to increase its contrast against a bright sky. The triangular gap in the round nebula is caused by a patch of dark nebula. You won't be able to see the dark nebula in amateur instruments, but you should be able to see a gap in the nebula's otherwise circular glow.

BLACK-EYE GALAXY (MESSIER 64)

BEAUTY: ★★✦ **BRAGGING RIGHTS:** A beautiful sight

HOW EASY IS IT TO SEE? Best with a rich-field telescope

BEST TIME TO SEE IT: Spring (in Coma Berenices)

TYPE: Galaxy **DISCOVERED:** 1779 by Edward Pigott

North Galactic Pole. The Milky Way forms a band of light across the sky. Since we are inside the Milky Way, it's hard to see galaxies behind all the stars and dust of the Milky Way's disk. But if you look away from the band of light, perpendicular to our galaxy's disk, you can see out into interstellar space: the realm of galaxies. Messier 64 is one of the brighter galaxies visible in that direction. In fact, it is very close to the North Galactic Pole (as seen from Earth)—the point directly above our galaxy's core. Find this galaxy on a line between Alpha and Gamma Comae Berenices. Photographs of M64 show a bright nucleus ringed by a dark semi-circle of black dust: it looks like an eye with a bruised lower lid. In a small telescope, however, you may not immediately see much detail. This is a subtle galaxy without sharp boundaries.

Patience, persistence, and practice will pay off.

⬤ *"I've seen it!"*

MESSIER 92

BEAUTY: ★★⌿ **BRAGGING RIGHTS:** A beautiful sight
HOW EASY IS IT TO SEE? Best with a high-power telescope
BEST TIME TO SEE IT: Summer (in Hercules)
TYPE: Globular Cluster **DISCOVERED:** 1777 by Johann Elert Bode

M13's sibling. Messier 92 doesn't get as much attention as its famous sibling, Messier 13 (the Hercules Cluster). Both are equally far from us, but M92 is both smaller and sparser than M13, making it less impressive. Nevertheless, it is one of the 10 brightest clusters in the Northern Hemisphere and well worth a look. Find it near Iota Herculis. Under extremely dark skies—a rarity today—M92 is just visible to the naked eye, but a good telescope will help you resolve this compact object. Start at 100× to examine the overall cluster, then increase your power to investigate its core. Averted vision, patience, and good skies are a must. With a little bit of imagination you might see dark lanes and channels crisscrossing Messier 92. What you're really seeing is the chance alignment of stars, perhaps accentuated by the sparseness of the cluster.

NGC 1333

BEAUTY: ★★⌐ **BRAGGING RIGHTS:** A beautiful sight
HOW EASY IS IT TO SEE? Telescope required
BEST TIME TO SEE IT: Winter (in Perseus)
TYPE: Diffuse Nebula **DISCOVERED:** 1858 by Eduard Schonfeld

Reflection nebula in Perseus. Many of the nebulae in this book are emission nebulae, in which newborn stars ionize their gas cocoons, which causes them to light up—it's the same principle behind neon signs. In contrast, NCG 1333 is a reflection nebula: a group of stars illuminates a patch of dust and gas—like headlights hitting fog.

Small and dim. NGC 1333 concentrates its meager light into a compact oval, making it easier to see. Nevertheless, this is not a simple object to spot: at low magnification it is too small to see details, but at high magnification, it is too dim. Look for it halfway between the Pleiades (page 76) and Algol (Beta Persei). Find it at low power first and increase magnification as appropriate.

"I've seen it!"

CALDWELL 53 (NGC 3115)

BEAUTY: ★★⁄ **BRAGGING RIGHTS:** A beautiful sight
HOW EASY IS IT TO SEE? Best with a rich-field telescope
BEST TIME TO SEE IT: Spring (in Sextans)
TYPE: Galaxy   **DISCOVERED:** 1787 by William Herschel

A lenticular galaxy. Caldwell 53 is a nearly edge-on galaxy, but if you compare it to NGC 4565 (page 170) you'll notice something missing: Caldwell 53 has no central dust lane! This is a lenticular galaxy—it's as featureless as an elliptical galaxy, but disk-shaped like a spiral galaxy.

Start from Lambda Hydrae. To find this galaxy, start at Lambda Hydrae, then move about 5 degrees north-northwest (see chart). In good, dark skies Caldwell 53 should be visible in binoculars. Its spindle-like shape will be visible even at low power.

X-ray source. The image above is a false-color rendering in which gold/yellow represents visible light and blue represents x-rays. The x-rays emitted from the core are evidence of a supermassive black hole.

MESSIER 46

BEAUTY: ★★✦ **BRAGGING RIGHTS:** A beautiful sight

HOW EASY IS IT TO SEE? Best with a small telescope

BEST TIME TO SEE IT: Winter (in Puppis)

TYPE: Open Cluster

DISCOVERED: Before 1654 by Giovanni Hodierna

Ghost cluster. This faint but fascinating cluster contains almost 200 faint stars. In binoculars it appears as a faint haze, but a telescope at 25× resolves it into a circular patch of tiny stars. True to its open cluster nature, there is no core or concentration—just stars vaguely aware of each other, like the faded memory of a once-great realm. Find this cluster about 5 degrees south of Alpha Monocerotis.

NGC 2438. With a good telescope and moderate magnification, you might see a ghostly circle just a little north of the cluster's center. This is NGC 2438, a planetary nebula that may or may not be associated with the cluster. Use averted vision and see if you can detect it.

○ *"I've seen it!"*

CALDWELL 13 (NGC 457)

BEAUTY: ★★☆   **BRAGGING RIGHTS:** A beautiful sight
HOW EASY IS IT TO SEE? Best with small telescope
BEST TIME TO SEE IT: Fall (in Cassiopeia)
TYPE: Open Cluster   **DISCOVERED:** 1787 by William Herschel

Best cluster in Cassiopeia? Caldwell 13 is sometimes overshadowed by nearby M103, but not for any logical reason: Caldwell 13 is both brighter and more interesting than its Messier rival. (It seems brand names count even in amateur astronomy.) Look for this cluster a couple of degrees south-southwest of Delta Cassiopeiae (see chart). Through binoculars it may look like a wisp of light around a relatively bright star. Ironically, it looks more like a comet than M103. (Messier compiled his original list of objects to avoid comet look-alikes, but he somehow missed this one.) At higher magnifications—maybe 75×—you'll resolve many of the stars in the cluster and perhaps start to see shapes. Do you see a stick figure? An airplane? Or maybe E.T.? All have been frequently reported by amateurs. What do you see?

MESSIER 19

BEAUTY: ★★⬩ **BRAGGING RIGHTS:** A beautiful sight
HOW EASY IS IT TO SEE? Best with a high-power telescope
BEST TIME TO SEE IT: Summer (in Ophiuchus)
TYPE: Globular Cluster   **DISCOVERED:** 1764 by Charles Messier

Elliptical Cluster. Messier 19 is visibly elongated even in small telescopes, making it stand out against its peers. M19 is closer than other clusters to the Milky Way's massive core; the tidal forces on it have stretched it out like soft taffy.

Start at moderate power. Find this cluster halfway between Antares and Theta Ophiuchi (see chart). You may not see more than a fuzzy shape, even at high power. M19's tight core makes it difficult to resolve, even with large instruments. Nevertheless, moderate power (about 100×) is sufficient to show its distinctive shape. Messier 19 is one of at least 10 globular clusters in Ophiuchus visible in small telescopes. M9, M10, M12, and M62, though not amazing enough to make the Top 101 list, are nevertheless noteworthy and worth a detour.

CALDWELL 22 (NGC 7662)

BEAUTY: ★★☽ **BRAGGING RIGHTS:** A beautiful sight
HOW EASY IS IT TO SEE? Best with a high-power telescope
BEST TIME TO SEE IT: Fall (in Andromeda)
TYPE: Planetary Nebula **DISCOVERED:** 1784 by William Herschel

Use high-power. Though bright enough to be seen with binoculars, low magnification won't show any detail of this planetary nebula. Find this nebula next to 13 Andromedae, then boost your magnification to 100× and admire its structural details.

Double-shelled nebula. Through a telescope you can clearly see the double-shell in Caldwell 22. You'll see a bright inner ring surrounded by a fainter, spherical glow. Unfortunately, the central star is invisible in moderate-size instruments. Though Caldwell 22 is known as the Blue Snowball Nebula, don't expect to see much color. Even in good skies and with powerful instruments you might not see much more than a pale bluish green color.

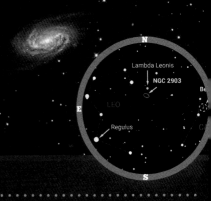

NGC 2903

BEAUTY: ★★⟩ BRAGGING RIGHTS: A beautiful sight
HOW EASY IS IT TO SEE? Best with a rich-field telescope
BEST TIME TO SEE IT: Spring (in Leo)
TYPE: Galaxy **DISCOVERED:** 1784 by William Herschel

Barred Spiral in Leo. Many years ago we thought the Milky Way galaxy looked like M74 (page 233): a galaxy in which spiral arms emerge from a central core. Instead, it now appears that we're a barred spiral, like this galaxy, NGC 2903. In the image above, a thin bar of stars and gas crosses the nucleus (roughly from 8 o'clock to 2 o'clock). The spiral arms emerge from this bar rather than from the center. To find this beautiful galaxy, look below the sickle of Leo. Your target is 1½ degrees south of Lambda Leonis. At low power it looks like a slightly fuzzy star. Using moderate power (75×) you might be able to see a couple of bright knots flanking the bright central core. These are the bright ends of the bar. Herschel thought he saw two bright nuclei (instead of one or three). What do you see?

MESSIER 94

BEAUTY: ★★✧   **BRAGGING RIGHTS:** A beautiful sight
HOW EASY IS IT TO SEE? Telescope required
BEST TIME TO SEE IT: Spring (in Canes Venatici)
TYPE: Galaxy   **DISCOVERED:** 1780 by Pierre Méchain

Galaxy in Canes Venatici. The tiny constellation of Canes Venatici boasts some of the best galaxies in the sky. We've already encountered M51 (page 168) and M106 (page 207); now we turn our attention to Messier 94. M94 is near the midpoint between the Alpha and Beta stars in the constellation.

A spiral galaxy? Messier 94 is classified as a spiral galaxy, but it may not look like one to you: there is no sign of distinctive arms like in M51. High-resolution astrophotos show dozens of tightly packed arms spiraling around the bright core. But to visual observers it looks more like a double-shelled planetary nebula. At 100× you'll see a bright ring around a tight core. Beyond the ring, the galaxy continues in a faint oval glow. Spend time examining the area around the core.

IRIS NEBULA (NGC 7023)

BEAUTY: ★★✦   **BRAGGING RIGHTS:** A beautiful sight
HOW EASY IS IT TO SEE? Telescope required
BEST TIME TO SEE IT: Fall (in Cepheus)
TYPE: Diffuse Nebula   **DISCOVERED:** 1794 by William Herschel

Nebula in Cepheus. This photogenic reflection nebula shines near the border of Cepheus and Draco, about 3½ degrees southwest of Beta Cephei. In binoculars, with averted vision, it looks like a fuzzy star, larger than those of comparable brightness.

Shrouded light. In astrophotos it looks like a purple flower full of folds, but to my eyes it is more like moonlight shining through a cloudy sky.

Knots and arcs. The central, bright part of the nebula is distinctly elongated, surrounded by a faint flow filled with subtle detail. At moderate power—say 100×—you might see subtle concentrations of light in even the fainter areas.

MESSIER 74

BEAUTY: ★★✦   **BRAGGING RIGHTS:** A beautiful sight
HOW EASY IS IT TO SEE? Best with a rich-field telescope
BEST TIME TO SEE IT: Fall (in Pisces)
TYPE: Galaxy   **DISCOVERED:** 1780 by Pierre Méchain

A beautiful spiral. Messier 74 is one of the most beautiful spiral galaxies—unfortunately, its charms are almost entirely confined to long-exposure astrophotos. Still, it is worth pursuing this sight visually, if only to challenge your skills. Your imagination will have to fill in the details.

Dark skies required. Wait for a crisp, moonless night in autumn, and find a location as far as possible from light pollution. South of the much brighter M33 (page 176) and a little over one degree away from Eta Piscium (see chart), you'll find the ghostly glow of Messier 74. Just seeing it is reward enough; many have tried and failed to spot this elusive target.

CERES

BEAUTY: ★⤸   **BRAGGING RIGHTS:** A beautiful sight
HOW EASY IS IT TO SEE? Telescope required
TYPE: Dwarf Planet   **DISCOVERED:** 1801 by Guiseppe Piazzi

Eleven planets. In the 1830s there were 11 planets in the solar system. Neptune and Pluto had yet to be discovered, but four "planets" were known between the orbits of Mars and Jupiter: Ceres, which was discovered first, plus Vesta, Juno, and Pallas. Eventually, though, dozens more were discovered in the same region, and astronomers quickly demoted them all to "asteroids." Today, Ceres is considered a dwarf planet, the largest member of the Main Belt of asteroids around the sun. The best time to see Ceres is during opposition—when it is opposite the sun as seen from Earth. At these times, Ceres is fully illuminated and as close as it can be to Earth. You'll see it as a sixth- or seventh-magnitude star. Though visible in binoculars, you'll need a good chart that plots the current position of Ceres among the background stars.

APPENDICES

CHOOSING BINOCULARS

There are many advantages to observing with binoculars instead of a telescope, and these include cost and portability. If you're just starting to observe the night sky, I highly recommend you start with binoculars.

Binoculars are rated using two numbers separated by an "×". The first is the *magnification* and the second is the *aperture*—the size of the front lenses—in millimeters. For example, 10×50 binoculars magnify 10 times and have front lenses 50 millimeters across.

7×35 binoculars are very common and suitable for astronomy. They have a wide field of view and enough aperture to reveal thousands of faint stars invisible to the naked eye. A view of the moon through these is wonderful, and bright objects such as the Orion Nebula and the Andromeda Galaxy are easily within reach.

More powerful binoculars can reveal more, but they come with trade-offs. For example, 15×70 binoculars, often advertised as "giant" binoculars, can show more detail in nebulae and star clusters, but they are heavy and difficult to hold steady. You will need a large tripod mount suitable for these giant binoculars.

If price is no object, consider buying image-stabilized binoculars. These electronic binoculars use a moving lens element to compensate for motion and vibrations.

CHOOSING A TELESCOPE

Choosing a telescope can be a daunting experience for beginners. There are so many models available, each with varying features and specs, that it's hard to know how to compare them. I often advise people to visit star parties, where they can look through different kinds of telescopes before buying their own.

Moreover, the best telescope for *you* depends on the kinds of objects you're interested in viewing. I would recommend one telescope for planets and a different one for galaxies and nebulae. In this section I'll describe the criteria you should pay attention to, and then follow up with a description of the most common kinds of telescopes.

MAGNIFICATION All telescopes magnify the image—that is, the view through a telescope appears larger than what the naked eye sees. 100x magnification means the image is 100 times larger.

Low magnification, generally from 7x to about 25x, is best for large, faint objects, like galaxies and nebulae. Though the object will appear smaller, all of its light will be concentrated in a small area, which will make it easier to see. High magnification, generally above 75x, is best for small, bright objects, like planets, globular clusters, and double stars. Remember, though, that high magnification requires calm, steady skies.

You can change the magnification of a telescope by using a different eyepiece. Larger eyepieces (25mm and above) provide lower magnification, while small eyepieces (10mm and below) provide higher magnification.

APERTURE, FOCAL LENGTH, AND FOCAL RATIO The most important spec is the *aperture* of the telescope. The aperture, expressed in either millimeters (for small scopes) or inches (for large scopes), measures the diameter of the *primary objective*—the main lens or mirror that gathers the faint light from all those distant objects.

All things being equal, a telescope with double the aperture has four times the light-gathering power. Stars in an 8-inch telescope will appear four times brighter than in a 4-inch telescope.

The *focal length* of a telescope, almost always expressed in millimeters, measures the distance between the primary objective and the focal point, generally where the eyepiece sits. The longer the distance, the greater the magnification. A telescope with twice the focal length magnifies twice as much, given the same eyepiece.

You might think higher magnification is better, but that's not always true. First of all, atmospheric distortion puts a limit on magnification. Unless the skies are pristine, magnifying more than 200-300× is a waste—the view will just get blurrier and blurrier. More importantly, when it comes to extended objects such as galaxies and nebulae, greater magnification makes the objects fainter.

It's easy to understand why. Your telescope has gathered a finite number of photons from some distant object, through the front of the telescope tube. Magnifying the image does not magically increase the number of photons—it just spreads them out over a larger area. The result is that each square millimeter of your eye gets fewer photons, and thus the image will seem fainter (though larger).

If larger apertures make things brighter and longer focal lengths make things fainter, then you need to consider both in order to determine how bright a nebula or galaxy will appear. Indeed it is the *ratio* of the two that determines the perceived brightness.

The *focal ratio* is the focal length divided by the aperture (in the same units). A 200-millimeter-diameter telescope (8 inches) with a 2,000-millimeter focal length has a focal ratio of 10, usually expressed as "f/10".

When considering galaxies and nebulae, smaller ("faster") focal ratios are better. An f/4 system is four times brighter than an f/8 system. Of course, this assumes the same eyepiece. You can increase brightness in a "slow" focal ratio telescope by using an eyepiece with less magnification.

For objects other than galaxies and nebulae, longer focal lengths are sometimes better. For example, planets and planetary nebulae are generally small and relatively bright. They are better viewed at higher magnifications, even if that dims the image somewhat. Similarly, stars are point-sources (not extended objects) and are not affected by magnification. They don't get much dimmer under magnification because they don't get bigger—they are too far away. Globular and open clusters are entirely composed of stars, and therefore benefit from magnification without much loss of brightness.

TELESCOPE MOUNTS After optics, the most important component of a telescope is the mount. A good, heavy mount keeps the image steady, even when the wind blows or when you accidentally brush against it. The worst mounts are those that shake for a long time at the slightest touch. No matter how good the optics, viewing under those conditions can be frustrating.

Telescope mounts often have clock drives to track the motion of the stars (actually, the Earth) so you don't have to constantly adjust the aim. The greater the magnification, the faster the object will move out of your field of view without a clock drive.

There are several different kinds of mounts. For the casual amateur, an inexpensive *alt-azimuth* mount is sufficient. For those interested in astrophotography, however, the more expensive *equatorial* mounts are a requirement.

Many telescope mounts have a built-in computer for pointing at specific objects in the sky. These "go-to" mounts can be very helpful for beginners, but only if properly aligned. Make sure the alignment procedure is easy and relatively accurate.

REFRACTOR TELESCOPES The first telescopes were all refractors. They are amazingly simple: a large lens at the front gathers the light and focuses it onto an eyepiece. Views through a refractor are sharp and high-contrast. Many people prefer them for viewing the planets.

A REFRACTOR TELESCOPE

Unfortunately, they have one major drawback: they are very expensive. For the price of a 3-inch refractor you could buy an 8-inch reflector of equivalent quality.

Most refractors are 70 to 120mm in aperture and f/6 to f/12. The faster ones are great for wide-field views of nebulae and large star clusters, while the slower ones are best for planets and small clusters.

I would not recommend a refractor as a beginner scope. The inexpensive ones are of low quality and seldom worth the price. But if you have experience, and are willing to pay the price, a good refractor can be an amazing instrument.

NEWTONIAN AND DOBSONIAN TELESCOPES
Sir Isaac Newton invented the reflecting telescope, which uses a curved mirror to focus light instead of a lens. The price advantage is immediately obvious: only one surface needs to be polished instead of two.

A second advantage is more subtle. Lenses suffer from *chromatic aberration*—different colors are fo-

A NEWTONIAN TELESCOPE

239

cused at different points—causing bright images to have color fringes. Correcting these distortions require more lenses—and more money. Reflecting telescopes don't suffer from chromatic aberration, and can thus be simpler.

But reflecting telescopes also have a drawback. The primary mirror reflects light and focuses it in front of it. You can't put your eye at the focal point because your head would block the telescope! Instead, a smaller secondary mirror bounces the light to the side where it can be focused by an eyepiece. But the secondary mirror blocks some of the light and causes a loss of contrast.

Nevertheless, the price advantages of a reflector are worth the trade-offs. All modern observatories, including the Hubble Space Telescope, use reflecting mirrors instead of lenses.

Newtonian telescopes come in all sizes, from tiny 4-inch aperture reflectors (like my old Astroscan) to monster 16-inch telescopes.

Larger telescopes are heavier and require a heavier (and more expensive) tripod mount. An alternative is the *Dobsonian* mount, popularized by John Dobson, which does away with the tripod and places the bottom of the telescope on a rotating "cannon mount".

A DOBSONIAN TELESCOPE

Dobsonian telescopes are very capable and probably the best option for apertures above a certain size. Their only drawback is that they cannot be used for long-exposure astrophotography.

Newtonian telescopes often have short focal-ratios—f/4 to f/6—and are ideal for faint nebulae and galaxies.

Bigger is often better, but remember that the best telescope is the one you actually use. If it's too hard to take your heavy telescope to a dark site, you will not use it as much. You might be better off with a smaller, more portable model.

SCHMIDT-CASSEGRAIN AND MAKSUTOV-CASSEGRAIN TELESCOPES An 8-inch aperture Newtonian telescope with an f/6 focal ratio is at least 48 inches long! A 16-inch one is at least 8 feet long—probably too large to fit in your car. Is there a better option?

The popular Schmidt- and Maksutov-Cassegrain telescopes were designed to have long focal lengths in a compact body. Light from the primary mirror bounces off the secondary, but instead of bending it to one side, it sends it back through a hole in the primary mirror. The result is a longer focal length in a smaller body.

Most Schmidt-Cassegrains are between 8 and 12 inches in aperture and have f/10 focal ratios. These specs make them good for planets, planetary nebulae, and small clusters. Faint nebulae and galaxies are not as bright as through a rich-field Newtonian, but still reasonable.

A MAKSUTOV CASSEGRAIN TELESCOPE

Maksutov-Cassegrains are usually smaller, around 4 inches in aperture, and have higher focal ratios: from f/12 to f/14. These are great for views of planets, lunar craters, and other small but bright objects.

HOW TO CHOOSE As with any other decision, experience makes things easier. The more time you spend looking through a telescope, the easier it will be for you to decide what you want. Star parties and local astronomy clubs (or just friends in the area) are your best bet for gaining experience without spending a lot of money.

With a little bit of familiarity, the choice will be easier.

ASTROPHOTOGRAPHY

Astrophotography is the hardest kind of photography. It takes a lot of equipment and a lot of practice to get even mediocre results. Lots of beginners quit because they expect to take an amazing photo on their first night.

You can take some pictures without a telescope. If you have DSLR with a good zoom lens you can start taking pictures of the night sky. You'll still need good image stacking software (see page 247) and you may want a tracking mount.

Unfortunately, the set of objects you can capture this way is small. Eventually you'll want more. Follow these steps if you want to try for more:

STEP 1: GET A TELESCOPE In astrophotography, the focal ratio is critically important (see page 237). In visual astronomy, you can use an eyepiece to change the magnification. But in astrophotography, there is no eyepiece. The camera is at *prime focus*, so the focal length of the telescope determines the magnification, and the focal ratio determines the brightness of the image.

If you want to take pictures of planets (which are small and bright), then you should get a high focal-ratio telescope (f/10 or higher). But if you want to take pictures of nebulae (which are large and dim), you need a "fast" ratio; f/4 is good if you can get it.

If necessary, you can use a focal reducer to decrease the focal ratio of a telescope. Or you can use a Barlow lens to increase the focal ratio.

STEP 2: GET A TRACKING MOUNT You need excellent optics, of course, but even more important is an accurate and steady mount. When you take pictures (particularly of faint galaxies) you'll need to expose a frame for up to an hour. That means your telescope has to exactly match the rotation of the Earth, or else your target will move in the frame and smudge the image. The more accurate the telescope mount, the easier this is.

Look for a tracking equatorial mount. Dobsonians are out, since they can't usually track. You should also avoid *alt-azimuth* mounts, which can track but cause image rotation. If at all possible, stick with a German Equatorial Mount or a fork mount with an equatorial wedge.

I recommend you spend at least 50 percent of your budget on a mount. Get a smaller telescope, if necessary; the mount will be worth it.

STEP 3: GET A CAMERA If you have a telescope and a tracking mount, you can point it at the moon and just take pictures with your phone camera! Even planets like Jupiter and Saturn can be captured by just pointing your phone camera at the eyepiece. As long as the object is bright enough, a quick exposure is enough. For fainter objects, though, you will need a camera directly connected to your telescope.

DSLR: You can do decent astrophotography with an off-the-shelf DSLR. All you need is the proper adapter to connect the camera to the telescope (T-adapter). These adapters are relatively cheap and all major cameras have them available.

The advantage of this path is that you might already have a DSLR, in which case you don't need to spend more money. I would definitely start here if that's the case.

The disadvantage is that DSLRs are not well-suited to astrophotography. For one thing, they all have an infrared filter over their sensor (because we don't want to see infrared in our pictures). For astronomy, unfortunately, that means we lose a good fraction of light.

Planetary Camera: If you're taking pictures of planets, then you need a fast, high-resolution camera. Planets are small and bright, so you don't have to worry about capturing a lot of light. Instead you need to worry about atmospheric distortion.

Planetary cameras take several shots per seconds, hoping to get one where the atmosphere is briefly steady. You end up with thousands of frames, from which you pick the best ones.

Fortunately, these kinds of cameras are cheap (~$200), so if you're interested in planetary photography, this is the way to go.

Monochrome Camera: Another disadvantage with DSLRs is that they take color pictures. Camera sensors just detect light intensity (not color), so DSLRs have color filters in a special pixel pattern (Bayer filter). The result is that some pixels are dedicated to red, some to green, and some to blue. When combined, we generate a full-color image. But one downside is that the resolution is reduced.

A different way to take color pictures is to take three complete images: one with a red filter, one with green, and one with blue. You can then combine the three channels to make a full-color image.

Another advantage is that you are not limited to color filters. You can use special narrowband filters, which focus on a very specific wavelength. This highlights the detail in nebulae. Most Hubble Space Telescope images, for example, are taken with narrowband filters. But if you get this far, you're definitely an advanced astrophotographer!

STEP 4: GET PROCESSING SOFTWARE Believe it or not, most of the time I spent on astrophotography was not on a telescope but in front of a computer.

The top frame is a single one-minute exposure of the galaxy M94. Note that a minute is actually a long time. In one minute the galaxy (actually, the Earth) has moved maybe a half of its diameter. If I didn't have a good mount you'd see a blurry streak.

The galaxy is very faint, just a little bit brighter than the surrounding background. We can enhance it by brightening the entire image—this is equivalent to the Levels or Curve command in Adobe Photoshop.

The second frame has enhanced the single exposure. Notice how noisy it is? This is just like taking a picture in very low light. We can't control the light of the galaxy, but we can expose for longer to reduce noise.

Unfortunately, you can't expose for too long. First, the more you expose, the greater the chance that your tracking will have errors, which will ruin the image. Second, if you expose too much, eventually the sky background will be so bright that the sensor gets saturated. Then you just have a solid white picture.

The answer is to take lots of shorter exposures and stack them together. Essentially, you mathematically add the two exposures to get a single exposure. The randomly distributed noise will cancel out, but the signal will get stronger! It sounds like magic, but it's quite common.

The third frame above is a composite of 220 individual exposures. Yes—you need that many to get a decent quality. Considering that I had to throw out many individual exposures due to tracking errors, the above required more than four hours of telescope time. And that's just for the monochrome version. Add more time to get the red, green, and blue images.

There is lots of stacking and processing software out there. If you buy a dedicated astrophotography camera, it usually comes with some software. Otherwise, I recommend something like Stark Lab's *Nebulosity*.

The final image combines the color stack and cleans up some of the background. For these kinds of tasks I use *Adobe Photoshop*, but any good image processing software will work.

STEP 5: GET A GUIDE CAMERA This step is optional. If you've gotten to step #4, then you're all set. You can take amazing images of a wide range of targets. But, eventually, you'll want to get a guide camera.

Remember that one of the biggest problems is accurate tracking. Even the best (i.e., most expensive) mounts have some amount of error. Gears and motors have tiny flaws.

To compensate, you need a second camera taking pictures of a single star, automatically communicating with the mount to adjust. When the guide camera sees the star move to the left, it tells the mount to slew to the right.

In order for this to work, you need a computerized mount that can receive commands in some way. You also need a way to mount the second camera so that it can see a guide start without getting in the way of your primary camera.

Astrophotography can be daunting, expensive, and frustrating, but the results are worth it. Give it a try!

BIBLIOGRAPHY

Chaikin, Andrew. *A Man on the Moon: The Voyages of the Apollo Astronauts*. New York: Penguin Books, 1994

Consolmagno, Guy and Davis, Dan M. *Turn Left at Orion*. Cambridge, United Kingdom: Cambridge University Press, 2003.

Dickinson, Terence. *Nightwatch: A Practical Guide to Viewing the Universe*: Buffalo, New York. Firefly Books, 1998.

Hoskin, Michael. *Discoverers of the Universe: William and Caroline Herschel*. Princeton, New Jersey: Princeton University Press, 2011.

O'Meara, Stephen James. *The Caldwell Objects*. Cambridge, Massachusetts: Sky Publishing Corporation, 2002.

O'Meara, Stephen James. *The Messier Objects, 2nd Edition*. New York: Cambridge University Press, 2014.

Poppele, Jonathan. *Night Sky: A Field Guide to the Constellation*s. Cambridge, Minnesota: Adventure Publications, 2009.

Sagan, Carl. *Cosmos*. New York: Random House, 1980.

Sagan, Carl. *Pale Blue Dot: A Vision of the Human Future in Space*. New York: Random House, 1994.

HELPFUL WEBSITES

Astronomy Magazine. www.astronomy.com.

Astronomy Picture of the Day. apod.nasa.gov/apod/astropix.html. An astronomy picture every day.

Sky & Telescope. www.skyandtelescope.com.

Stellarium. www.stellarium.org. Free, open source planetarium software.

GLOSSARY

accretion disk: A superheated disk of dust and gas spiraling into a high-gravity object such as a star, white dwarf, or black hole. The intense energies of the disk cause it to emit electromagnetic radiation.

active galaxy: A galaxy that is emitting intense electromagnetic radiation out of its nucleus, generally due to matter falling into a supermassive black hole.

alt-azimuth mount: A type of telescope mount in which the axes are aligned to the horizon and the **zenith**. The *altitude* axis points the telescope up and down, to any point above the horizon. The *azimuth* axis rotates the telescope to any compass angle. Alt-azimuth mounts must move on both axes to compensate for the Earth's rotation.

annular eclipse: An eclipse of the sun in which the moon fits inside the disk of the sun, leaving a bright, ring of sunlight.

aperture: The diameter of a telescope's primary mirror or lens. Larger diameters are able to gather more light and have better resolving power.

arcminute: A measure of angular distance between two points in the sky. There are 180 degrees of sky from the eastern horizon, through the zenith, to the western horizon. There are 60 arcminutes in a degree. The full moon is approximately 30 arcminutes in diameter.

arcsecond: A measure of angular distance equal to $1/60$ of an arcminute.

averted vision: An observing technique in which you look at a faint object out of the corner of your eye, instead of directly at it. Peripheral vision is more sensitive to light than direct vision.

Big Bang: The rapid expansion of space at the beginning of the universe, which explains why most galaxies seem to be moving away from each other.

Caldwell Catalog: A catalog of beautiful, non-Messier, deep-sky objects for amateur astronomers to observe. Compiled by Sir Patrick Caldwell Moore in 1995. See page 152.

Cepheid variable: A type of variable star used to determine true distances to galaxies and other objects. See page 176.

circumpolar: An object that never sets because it is close to a celestial pole (as seen from the observer). The Little Dipper, for example, is circumpolar.

comet: A small, icy body in the solar system that heats up and emits gases when its orbit brings it close to the sun. See page 48.

conjunction: An event in which two or more objects appear close to each other in the sky; for example, a conjunction between Jupiter and Mars.

constellation: A region of the celestial sphere, defined by the International Astronomical Union, and based on traditional patterns of stars.

cosmos: All that is or ever was or ever will be. A synonym for the universe.

degree: A measure of angular distance in the sky. There are 180 degrees between the eastern horizon and the western horizon. There are 60 **arcminutes** in a degree and 60 arcseconds in an arcminute.

eclipse: An event in which one object passes in front of the sun, casting a shadow on another object. When the moon passes in front of the sun, it is a solar eclipse. When the Earth passes in front of the sun, as seen from the moon, it is a lunar eclipse.

ecliptic: The apparent path of the sun on the celestial sphere, due to the plane of Earth's orbit. All planets orbit around the sun close to the same plane, and thus the planets all appear near the ecliptic.

electromagnetic radiation: A type of energy, consisting of photons, and emitted by various processes at various wavelengths. Almost everything we know about objects in space comes from detecting and focusing electromagnetic radiation in all its forms: radio waves, microwaves, infrared light, visible light, ultraviolet light, x-rays, and gamma rays.

equatorial mount: A type of telescope mount that is aligned with the Earth's axis of rotation. Equatorial mounts can compensate for the Earth's rotation by counter-rotating on a single axis, making them ideal for astrophotography.

eyepiece: A removable lens that you look through on a telescope. Different eyepieces provide different magnification levels.

filter: An optical component, added to a telescope, to exclude certain frequencies of light. Some filters can exclude light pollution (to enhance the view of nebulae); others can exclude all but certain colors, to enhance the view of planets.

focal length: The length of a telescope's focal point. Longer focal lengths have greater magnification.

focal ratio: The ratio of a telescope's **focal length** to its aperture. For example, an f/4 telescope has a focal length four times longer than its aperture. Shorter focal ratios make extended objects (like galaxies and nebulae) appear brighter.

galaxy: An enormous assemblage of gravitationally bound stars, gas, and dust. The Milky Way is a galaxy. There are billions of galaxies in the universe, each containing from billions to trillions of stars.

globular cluster: A spherical cluster of stars, generally found away from the galactic disk.

ionization: A process in which an atom gains or loses an electron, giving it a negative or positive charge. Nebulae, which are often composed of ionized gas, emit visible light.

Kuiper Belt: A disk of asteroids and comets orbiting beyond Neptune. Pluto is the largest member of the Kuiper Belt.

light pollution: The brightening of the night sky caused by artificial lights. Light pollution decreases the contrast of faint objects like galaxies and nebulae, making it harder to observe them. Light pollution can sometimes be mitigated by filters, but the best solution is to observe in National Parks and other areas far from city lights.

magnification: The scaling up of a visual image in a telescope. Higher magnification can reveal more detail in small objects (such as planetary nebulae), but it also makes objects dimmer, which is undesirable for faint objects such as galaxies. The amount of magnification depends on the telescope's **focal length** and the **eyepiece** being used.

magnitude: A measure of the brightness of a star or other object. The star Vega is magnitude 0. Objects fainter than Vega have higher magnitude numbers; objects brighter than Vega have negative magnitude numbers. In dark skies, the naked eye can see stars down to magnitude 6.

mare: A large, dark plain on Earth's moon. Early astronomers thought they might be seas, and named them *maria*, which is Latin for sea.

Messier Catalog: A catalog of beautiful deep-sky objects compiled by comet-hunter Charles Messier as a means to avoid objects that could be confused for comets.

NGC: New General Catalog. A comprehensive catalog of deep-sky objects compiled by John Dreyer in 1888. Most of the brightest deep-sky objects have an NGC designation.

nebula: A vast cloud of interstellar gas and dust. *Diffuse nebulae* are extended and have ill-defined boundaries; some diffuse nebulae give birth to entire clusters of stars. *Planetary nebulae* are the remains of a red giant star after it has collapsed into a white dwarf.

occultation: An event in which an object passes in front of another, temporarily obscuring it. For example, the moon can *occlude* Mars when it passes in front of it (as seen from Earth).

open cluster: A loose association of stars, generally born in the same nebula, and still moving together in the galaxy. Open clusters are only loosely bound by gravity, and eventually disperse.

opposition: The point at which a planet is opposite the sun as seen from Earth. This is one of the best times to see the planet, since it is fully illuminated, and closer to Earth than at other points in its orbit. Inner planets (Mercury and Venus) are never at opposition.

parallax: The difference in the apparent position of an object when seen from two slightly different points, which can be used to determine true distance. The nearest stars exhibit parallax when seen from opposite ends of Earth's orbit (e.g., July vs. December), which can be used to estimate the star's distance.

pareidolia: A phenomenon in which the brain sees a familiar object from a random arrangement of shapes. For example, seeing objects in an ink blot or faces on a rock cliff.

penumbra: In an **eclipse**, the region in which an object is only partially in shadow. See also, **umbra**.

photon: A fundamental particle which makes up all **electromagnetic radiation**, including visible light.

prominence: A solar phenomenon in which bright gaseous features extend outward from the sun's surface. Prominences may be visible during a solar eclipse or with suitable filters.

quasar: A quasi-stellar radio source; the most distant and energetic kind of active galaxy.

reflecting telescope: A kind of telescope using a curved mirror as the primary means to collect and focus light.

refracting telescope: A kind of telescope using a lens as the primary means to collect and focus light.

retrograde: The apparent backwards motion of certain planets as seen from Earth. Mars, for example, exhibits retrograde motion when Earth passes it in its orbit.

rich-field telescope: A kind of telescope with short focal ratios, general f/5 or better. Rich-field telescopes are ideal for observing nebulae and galaxies.

spectrum: The range of light at various wavelengths. By measuring the intensity of an object's light at various wavelengths, we can learn about its physical composition. For example, nebulae emit light at the same wavelength as ionized hydrogen, suggesting that they are mostly composed of hydrogen.

sunspot: A cool spot on the sun's surface caused by concentrations of magnetic field lines.

supernova: A cataclysmic release of energy when a star runs out of fuel and collapses due to its own gravity.

supernova remnant: The scattered remains of a supernova explosion, generally consisting of dust and gas, expanding away from the star.

terminator: The line on the moon's surface dividing the sunlit side from the shadowed side. Lunar features such as craters and mountains are easily visible at the terminator because their shadows are long.

turbulence: The random motion of the atmosphere causing stars to twinkle. Turbulence limits the amount of telescopic magnification that can be used.

umbra: In an eclipse, the region in which an object is completely in shadow. See also, **penumbra**.

wide-field telescope: See **rich-field telescope**.

zenith: The point in the sky directly overhead from the observer's point of view.

CHECKLIST

ABOUT THE AUTHOR

George Moromisato is an American software engineer, game designer, and astrophotographer. He's been programming for 70 percent of his life and has worked at Microsoft, IBM, and Lotus Development Corporation. He is the designer of several computer games, including *Anacreon* (one of the first 4X games), *Chron X* (the first online collectible card game), and most recently, *Transcendence*. But his first love is astronomy, and he avidly photographs the night sky whenever clouds allow. He is the co-founder of PhotonSky.com, an upcoming site for sharing astrophotos and promoting amateur astronomy. You can reach him via email: george.moromisato@photonsky.com.